A PLUME BOOK

SPYCHIPS

DR. KATHERINE ALBRECHT is the founder and director of CASPIAN, an international consumer group. Dubbed the "Erin Brockovich" of RFID by *Wired* magazine, she is one of the leading voices for privacy in today's fast-changing, high-tech world. Katherine holds a doctorate in education from Harvard University.

LIZ MCINTYRE is an award-winning investigative writer with a flair for exposing corporate shenanigans and bureaucratic misdeeds. She serves as CASPIAN's communications director, and has been the master strategist for many of the organization's most successful media campaigns.

Praise for *Spychips*

"The book makes a very persuasive case that some of America's biggest companies want to embed tracking technology into virtually everything we own, and then study our usage patterns twenty-four hours a day. It's a truly creepy book and well worth reading."
— Hiawatha Bray, technology reporter, *The Boston Globe*

"Brilliantly written . . . so full of fascinating vignettes and facts that I can't put it down."
— Claire Wolfe, author of *Don't Shoot the Bastards (Yet): 101 More Ways to Salvage Freedom*

"*Spychips* is one of the best privacy books in many years. . . . The privacy movement needs a book. I nominate *Spychips.*"
— Marc Rotenberg, executive director, the Electronic Privacy Information Center (EPIC)

"The ultimate privacy argument against RFID . . . This won't be comfortable reading in the IT departments of major retailers and manufacturers, but it is essential."
— Evan Schuman, *CIO Insight*

SPYCHIPS™

> ▶▶ HOW MAJOR CORPORATIONS AND GOVERNMENT PLAN
> TO TRACK YOUR EVERY PURCHASE AND WATCH
> YOUR EVERY MOVE ◀◀

KATHERINE ALBRECHT AND LIZ MCINTYRE

A PLUME BOOK

PLUME
Published by Penguin Group
Penguin Group (USA) Inc., 375 Hudson Street, New York, New York 10014, U.S.A.
Penguin Group (Canada), 90 Eglinton Avenue East, Suite 700, Toronto, Ontario, Canada M4P 2Y3
(a division of Pearson Penguin Canada Inc.)
Penguin Books Ltd., 80 Strand, London WC2R 0RL, England
Penguin Ireland, 25 St. Stephen's Green, Dublin 2, Ireland (a division of Penguin Books Ltd.)
Penguin Group (Australia), 250 Camberwell Road, Camberwell, Victoria 3124, Australia
(a division of Pearson Australia Group Pty. Ltd.)
Penguin Books India Pvt. Ltd., 11 Community Centre, Panchsheel Park, New Delhi – 110 017, India
Penguin Books (NZ), cnr Airborne and Rosedale Roads, Albany, Auckland 1310, New Zealand
(a division of Pearson New Zealand Ltd.)
Penguin Books (South Africa) (Pty.) Ltd., 24 Sturdee Avenue, Rosebank, Johannesburg 2196, South Africa

Penguin Books Ltd., Registered Offices: 80 Strand, London WC2R 0RL, England

Published by Plume, a member of Penguin Group (USA) Inc. This is an authorized reprint of a hardcover edition
published by Nelson Current, a division of Thomas Nelson, Inc. For information address
SpecialMarkets@ThomasNelson.com.

First Plume Printing, October 2006
10 9 8 7 6 5 4 3 2 1

Spychips is a trademark of Katherine Albrecht and Liz McIntyre

 REGISTERED TRADEMARK—MARCA REGISTRADA

The Library of Congress has catalogued the Nelson Current edition as follows:

Albrecht, Katherine.
Spychips : how major corporations and government plan to track your every move
    with RFID / Katherine Albrecht and Liz McIntyre.
    p. cm.
    Includes index.
ISBN 1-5955-5020-8 (hc.)
ISBN 0-452-28766-9 (pbk.)
    1. Privacy, Right of—United States.   2. Radio frequency identification systems.
3. Electronic surveillance—Social aspects—United States.   I. Title: Spy chips.
II. Title: How government and major corporations are tracking your every move.
III. McIntyre, Liz.   IV. Title.

323,44'82—dc22   2005020104

Printed in the United States of America

We dedicate this book to the millions of people who have worked to fight oppression throughout history. Special thanks to the members of CASPIAN who recognized the RFID threat and helped us sound the alarm.

This book would have been impossible without the love and support of our husbands and children. Thank you for tolerating the late nights and marginal dinners. We'll make it up to you. And special thanks to our mothers who taught us to believe in ourselves and stand up for what's right, even when it meant breaking ranks with our peers. You're every bit a part of this book, and it couldn't have been written without you.

Love,

Katherine and Liz

CONTENTS

▶▶ ◀◀

▶▶　◀◀

We all want progress . . . [but] if you're on the wrong road, progress means doing an about-turn and walking back to the right road; and in that case, the man who turns back soonest is the most progressive man. . . . We are on the wrong road. And if that is so, we must go back. Going back is the quickest way on.

—C.S. Lewis

FOREWORD

[by Bruce Sterling]

Everybody has a role in the RFID industry, because, as this remarkable book makes clear, we're not offered any choice about it. If you've never heard of RFIDs or "spychips," it would be quite a good idea to read this book pretty soon. It's very topical.

If you have any direct role within the RFID industry, then you need to read this book instantly. Hurry. Waste not another precious moment. You won't like this book. *Spychips* will hurt your feelings. You will blush, and itch, and sweat, and drum your heels, and perhaps tear entire chapters out with squalls of rage, to see a work about your industry that is so jaundiced, and uncharitable, and unflinchingly suspicious, and that makes so much effective, highly damaging, public fun at your expense. So read it, and make all your coworkers read it. You will learn a host of painful, valuable things in a hurry. For you, it may not yet be too late.

There have been many tech manuals and white papers written about RFIDs. They're mostly quite technical, all about transponders and supply chains and megahertz, with maybe a few bits about value-adding and stakeholder value. I'm a tech journalist, so I read a lot of dry, boring crap like that.

But RFID is not high-tech or hard to understand. It is not confusing, sophisticated, or arcane. RFID is very dumb computer tech, the kind of computer tech
that even grocers can understand. There's no need to get intimidated by the
technology—because the issues here are all about money and power.

This book is the most exciting book about RFID ever written. This is the one
RFID book that every RFID enthusiast must own. Not because the book is
enthusiastic about the new technology—but because it's full of passionate,
stinging contempt. It's like watching Big Brother come home and get a rolling
pin broken over his head by Mrs. Big Brother, who knows that, even though he
thinks he's everybody's daddy, he's a stalker, and a voyeur, and a crook, and a
cheat, and drunk on his own ego, and a handwashing, sniveling deadbeat who
ought to be ashamed of himself.

This is the Devil's Dictionary for RFID, and in its own dainty, feminine,
rapier-tongued way, this is a masterpiece of technocriticism. The nascent RFID
industry is not Big Brother. Not yet, anyhow. Instead, it is a giant toddler whose
supermarket diapers are already richly soiled. It's sure got a mighty ton of dirty
laundry for a baby still that small, and in Katherine Albrecht and Liz McIntyre,
the RFID industry has found a hardworking pair who'll willingly scrub that
laundry, name and number every stain, and then pin it out to dry.

These two unique individuals, the Lone Ranger and Tonto of the RFID
frontier, are the nightmare scenario for the computerized retail superstore of
tomorrow, because they're the computerized super female consumer advocates
of tomorrow. And boy have they ever got their industry's number. They've got
all two-to-the-ninety-sixth-power digits of it.

To understand what species of book this is, let me offer a historical analogy.
Imagine yourself cruising along in the 1950s chemical industry, happily
patenting and spreading potent toxins. Then, this searching, thoughtful female
journalist, Rachel Carson, who doesn't even have a chemistry degree, comes out
of nowhere. A classic popular muckraker, Ms. Carson points out to a shocked
public that you're killing not just the mosquitoes but all the pretty butterflies
and birds. She writes *Silent Spring*, and it's so influential and damning that even
your own kids decide you must be nuts. That's also what's happening here.

To its credit, the RFID industry is very twenty-first century, and therefore a little cagier than the pesticide biz in the 1950s. Realizing that they had a world-shattering technical breakthrough at hand, they hired a top-notch public relations firm first to go fish in the waters of public acceptance. Acceptance of what, exactly? Basically, acceptance of what this book describes in detail: an amazingly ambitious scheme to infest the entire physical infrastructure of the planet with a spray-on global blanket of Internet interactivity. This is truly a fabulous, earth-shaking scheme. It is awesome.

The hired PR firm, the gifted Fleishman-Hillard outfit, poked around some, trying to tell everyday people what this huge revolution might mean to them.

Having briefed a few focus groups, the PR guys returned and told the industry's founders that normal consumers would surely react with superstitious horror and unfeigned Luddite dread. That wasn't good news. But the promised rewards were colossal, so the techies waded in anyway. They decided that the public should be told as little as possible about their project. Whatever the public learned should be obfuscated as much as possible, until the installation of RFID worldwide had become a *fait accompli*. So, first it would be obscure; then it would be old hat; and, with any luck, it would never quite become a public issue at all. But, well, there's a big hitch. That's the so-called "secrecy." The Internet of Things is supposed to be invisible to all but its corporate and military masters. But the Internet itself is hugely obvious and famous—because, even though the Internet is also corporate and military in its origins, for about a decade the Internet was all anybody ever talked about. You can't possibly have a hugely famous Internet made of pixels and an ultra-quiet Internet made of actual consumer objects. So we're seeing a violent collision of two models here: two loud, flamboyant, irrepressible Internet activists, researching and publicizing the secretive, business-confidential Internet of Things.

Anybody who can create that link between the worlds is gonna get justly famous, and Katherine Albrecht (judging by Google and the hundreds of journalists she has briefed) is already, by far, the most famous RFID expert in the whole wide world. She thinks RFID is an evil crock, but she's sure got a lot to say about it—all of it is fascinating, some is gross and revolting, and most of it is hilarious. This is the first, and maybe the loudest, popular book on a crucial technology of our times. It's not the full or final story—it's a futurist book, in anticipation of the story—but history will treat this book kindly.

As this book demonstrates irrefutably, the RFID industry has patented some fantastically sinister, sci-fi style business notions. The authors are not making these things up—the industry is. Patents are public documents, not trade secrets. Anybody can go look at patents. It's just, well, somehow, nobody was ever supposed to notice them or care.

Why? Because this is an industry with some deeply schizoid doublethink problems, which come directly from its wacky origins in the spy and security communities.

The people of the RFID biz are very covert, spooky, and security-conscious, with deep, profitable ties into Homeland Security and the Pentagon. And yet, they're also very large, everyday public companies: Wal-Mart, Procter & Gamble, Tesco, Benetton, Philips, IBM, Cisco, Exxon-Mobil—dozens of familiar, everyday, publicly-traded companies with big, soft, squishy, publicity-conscious brands.

It's really hard to be a big, public, for-profit spy with tons of shareholders, zillions of customers, and even employees who don't like you very much. That scheme doesn't hang together. Riddle me this: How do you profit by telling your own shareholders that you've bugged their own clothes and shoes with tiny radio transponders from your stores? How can you have a board meeting when the clothes and shoes of your own board members might be full of a competitor's spychips? These eager pioneers have failed to think these issues through, mostly because they never expected or planned to face a reality check. But their situation is inherently unstable.

Enter Katherine Albrecht with her red suit, red hair, a TV talking head's makeover, and mirrored sunglasses. Still a university student, she places the new surveillance industry under some mild doctoral-dissertation surveillance of her own, and is astounded. She finds in short order that she can win awestruck, worldwide press attention just by repeating the industry's own private pep talks in public. She becomes the instant, planetary, go-to expert on RFID—mostly because the real experts on RFID are so anxious to keep mum.

There's no need to unravel a Watergate break-in here; the so-called "secret" is literally and physically scattered all over the landscape. RFID bugs are attached to diaper boxes, shampoo bottles, and women's underwear, and they cost a few cents each and are supposed to become ubiquitous. Only nobody is supposed to notice or care. *Huh?* All you have to do is point at the emperor's RFIDs; it's like revealing lice in the royal gown.

This book is a comprehensive, detailed, and footnoted work of corporate futurism. But, unlike most such futurist works, it's not saccharine industry boosterism. *Spychips* is something new in the corporate world: it's the work of Early Dys-Adopters, of Power-Unusers, of online activists who fully understand promotion, marketing, and effective PR and then use new media tools to beat unwise companies into pulps instead of serving as their paid handmaidens.

The authors of this book lack big budgets, a power base, or an agenda. They are, however, energetic, clever, highly motivated, highly wired, and chock-full of feminine wiles. Thanks mostly due to legwork, Google, and chatty e-mail from many like-minded souls, they have become a retailer's worst nightmare. They are as uncontainable and global as the industry they decry, for they are the Digitized Suburban Mom Shoppers from Hell: perceptive, well-connected, entirely self-educated, very American, highly skilled industry gurus; quotable, word-of-mouth branding killers with viral marketing voodoo; digital Cassandras who are second to none in downsides, dirty laundry, and doomsaying. Plus, they are witty and good-looking.

I expect to spend the next ten years watching the next Internet Revolution— but the New Grocers of the Internet of Things have already gotten the customers they deserve.

SPYCHIPS™

1

Tracking Everything Everywhere

▶▶ The RFID Threat ◀◀

RFID will have a pervasive impact on every aspect of civilization, much the same way the printing press, the industrial revolution and the Internet and personal computers have transformed society. . . . RFID is a big deal. Its impact will be pervasive, personal and profound. It will be the biggest deal since Edison gave us the light bulb.

> —Rick Duris,
> *Frontline Solutions Magazine*, December 2003[1]

Technology . . . is a queer thing. It brings you great gifts with one hand, and it stabs you in the back with the other.

> —C.P. Snow, *New York Times*, 1971[2]

Imagine a world of no more privacy.

Where your every purchase is monitored and recorded in a database and your every belonging is numbered. Where someone many states away or perhaps in another country has a record of everything you have ever bought, of everything you have ever owned, of every item of clothing in your closet—every pair of shoes. What's more, these items can even be tracked remotely.

Once your every possession is recorded in a database and can be tracked, you can also be tracked and monitored remotely through the things you wear, carry, and interact with every day.

We may be standing on the brink of that terrifying world if global corporations and government agencies have their way. It's the world that Wal-Mart,

Target, Gillette, Procter & Gamble, Kraft, IBM, and even the United States Postal Service want to usher in within the next ten years.

It's the world of radio frequency identification.

Radio frequency identification, RFID for short, is a technology that uses tiny computer chips—some smaller than a grain of sand—to track items at a distance. If the master planners have their way, every object—from shoes to cars—will carry one of these tiny computer chips that can be used to spy on you without your knowledge or consent. We've nicknamed these tiny devices "spychips" because of their surveillance potential.

> ▶▶   "THE PRIVACY IMPACT OF LETTING
> MANUFACTURERS AND STORES PUT RFID
> CHIPS IN THE CLOTHES, GROCERIES, AND
> EVERYTHING ELSE YOU BUY IS ENORMOUS."
> —CALIFORNIA STATE SENATOR
> DEBRA BOWEN[3] ◀◀

If you've been staying in touch with the news about RFID, you may already know who we are and something of the public battles we have fought to try to keep this technology off of consumer products and out of our homes. In case you don't know who we are and why we can make such claims with conviction, an introduction is in order.

We are Katherine Albrecht, founder and director of CASPIAN (Consumers Against Supermarket Privacy Invasion and Numbering), and Liz McIntyre, the organization's communications director. CASPIAN is a grass-roots organization that has been tackling consumer privacy issues since 1999.* In the pages that follow, we'll give you a ringside seat to some of the battles we've fought with companies like Benetton, Gillette, and retail giant Tesco. You'll see why

---

* With close to fifteen thousand members in all fifty U.S. states and over thirty countries worldwide, CASPIAN seeks to educate consumers about marketing strategies that invade their privacy and to encourage privacy-conscious shopping habits across the retail spectrum.

Advertising Age says our presence has been felt from Berlin to Bentonville (corporate home of Wal-Mart), and you'll also learn how we uncovered plans by companies to track consumers around stores, use RFID to spam consumers with personalized advertising, and even monitor what people do in their own homes.

We're also suburban moms who've taken on some of the largest corporations in the world because we care about the future our children will inherit if this dangerous technology is unopposed. We believe consumers should know what's in store so we can work together to protect our privacy and civil liberties before it's too late.

We know that a Big Brother vision of the future sounds farfetched. We didn't believe it ourselves until we saw with our own eyes and heard with our own ears companies detailing their mind-boggling plans. We assure you that this seemingly impossible future is on the drawing board, and we promise that by the time you finish this book, you will be convinced, too.

For the past several years, we have devoted ourselves full-time to combing every article, reading every white paper, pursuing every insider tip, and scanning through thousands of patent documents to piece together a picture of this planned RFID future. We've attended trade shows, sat in on top level meetings, and had long talks with the people implementing these plans.

What we learned will shock you.

If anything you read in the following pages strikes you as improbable, please refer to the endnotes at the back of the book. We've included hundreds of references to original source materials that should satisfy even the most skeptical reader.

In a future world laced with RFID spychips, cards in your wallet could "squeal" on you as you enter malls, retail outlets, and grocery stores, announcing your presence and value to businesses. Reader devices hidden in the doors, walls, displays, and floors could frisk the RFID chips in your clothes and other items on your person to determine your age, sex, and preferences. Since spychip information travels through clothing, they could even get a peek at the color and size of your underwear.

We're not joking. A major worldwide clothing manufacturer named Benetton has already tried to embed RFID chips into women's undergarments. And they would have gotten away with it, too, had it not been for an international outcry when we exposed their plan. Details of the "I'd Rather Go Naked" campaign come later in the book.

While consumers might be able to avoid spychipped clothing brands for now, they could be forced to wear RFID-enabled work clothes to earn a living. Already uniform companies like AmeriPride and Cintas are embedding RFID tracking tags into their clothes that can withstand high temperature commercial washings.

Don't have to wear a chipped uniform to work? Your RFID-enabled employee badge could do the spying instead. One day, these devices could tell management whom you're chatting with at the water cooler and how long you've spent in the restroom—even whether or not you've washed your hands.

Our next generation of workers could be conditioned to obediently accept this degrading surveillance through forced early exposure. Some schools are already requiring students to wear spychipped identification badges around their necks to keep closer tabs on their daily activities. If Johnny is one minute late for math class, the system knows. It's always watching.

Retailers are thrilled at the idea of being able to price products according to your purchase history and value to the store. RFID will allow them to assess your worth as you pick up products and flash you a corresponding customer-specific price. Prime customers might pay three dollars for a staple like peanut butter while "bargain shoppers" or the economically challenged could be charged twice as much. The goal is to encourage the loyalty of shoppers who contribute to the profit margins while discouraging those who don't. After all, stores justify, why have unprofitable customers cluttering the store and breathing their air?

RFID chips embedded in passbooks and ATM cards will identify and profile customers as they enter bank lobbies, beaming bank balances to employees who will snicker at the customer with a mere thirty-seven dollars in the bank while offering white glove treatment to the high-rollers.

RFID could also be used to infringe upon civil liberties. The technology could give government officials the ability to electronically frisk citizens without their knowledge and set up invisible checkpoints on roads and in pedestrian zones to monitor their movements.

While RFID proponents claim they would never use RFID to track people, we will prove they are not only considering it, they've done it. The United States government has already controlled people with RFID-laced bracelets—and not just criminals. And now they've begun embedding spychips in U.S. passports and travel documents so travelers can be tracked as they cross international borders.

Hitting the open road will no longer be the "get away from it all" experience many of us crave. You may already be under surveillance, courtesy of your RFID-enabled highway toll transponder. Some highways, like those in the Houston area, have set up readers that probe the tag's information every few miles. But that's child's play compared to what they've got planned. The Federal Highway Administration is joining with states and vehicle manufacturers to promote "intelligent vehicles" that can be monitored and tracked through built-in RFID devices (*Minority Report*-style).

RFID spychips in your shoes and car tires will make it possible for strangers to track you as you walk and drive through public and private spaces, betraying your habits and the deepest secrets even your own mother has no right knowing. Pair RFID devices with global positioning (GPS) technology, and you could literally be pinpointed on the globe in real time, creating a borderless tracking system that already has law enforcement, governments, stalkers, and voyeurs salivating.

There will be no more secret love letters in the RFID world, either—not if the U.S. Postal Service has its way. They would like to embed every postage stamp with an RFID chip that would enable point-to-point tracking. Even more disturbingly, RFID could remove the anonymity of cash. Already, the European Union has discussed chipping Euro banknotes, and the Bank of Japan is contemplating a similar program for high-value currency. Your every purchase could be under the microscope.

So could your trash. In the RFID world, garbage will become a snoop's and criminal's best friend. Today, it's a dirty job sifting through diapers and table scraps to get at tell-tale signs of a household's market value, habits, and purchases. In the RFID world, scanning trash could be as simple as driving down the street with a car-mounted reader on trash day.

How about the "smart" house? Researchers have developed prototype "homes of the future" to showcase RFID-enabled household gadgets like refrigerators that know what's in them (and can tattletale to marketers), medicine cabinets that talk (to your doctor, government, and HMO), and floors that keep track of where you are at each moment. The potential is staggering. Your insurance company could remotely monitor your food consumption and set rates accordingly, health officials could track the prescription drugs you're taking, and attorneys could subpoena your home activity records for use against you in court.

Home RFID networks will allow family members to remotely track you during your "golden years," or times of incompetence, real or otherwise. Doors can remain bolted to keep you from wandering, toilets can monitor your bowel habits and transmit data to distant physicians, and databases can sense your state of mind. It's all under development and headed your way.

But chipping inanimate objects is just the start. The endpoint is a form of RFID that can be injected into flesh. Pets and livestock are already being chipped, and there are those who believe humans should be next. Incredibly, bars have begun implanting their patrons with glass-encapsulated RFID tags that can be used to pay for drinks. This application startles many Christians who have likened payment applications of RFID to biblical predictions about the Mark of the Beast, a number the book of Revelation says will be needed to buy or sell in the "end times."

While some of these applications are slated for our future, others are already here, right now—and they're spreading. Wal-Mart has mandated that its top suppliers affix RFID tags to crates and pallets being shipped to selected warehouses. Analysts estimate this one initiative alone has already driven hundreds of millions of dollars' worth of investment into the technology.[4]

Other retailers such as Albertsons, Target, and Best Buy have followed suit with mandates of their own. According to one industry analyst, in 2004 there were already sixty thousand companies operating under RFID mandates and scrambling to get with the spychip program as quickly as possible.[5]

Adding fuel to the fire, the Department of Defense is also requiring suppliers to use RFID. In fact, government cheerleaders can't fall over themselves fast enough to support the technology. The Department of Homeland Security is testing the use of RFID in visas, and the Social Security Administration is using spychips to track citizen files. Not to be outdone, the Food and Drug Administration wants RFID on all prescription drugs, and the makers of Oxycontin and Viagra have already begun to comply. The FDA has also approved the use of subcutaneous RFID implants for managing patient medical records—the same implants being used to track bar patrons.

You may have already brought a spychip home with you. If you own a Mobil Speedpass, you're interacting with RFID every time you use it. And if you bought Procter & Gamble's Lipfinity lipstick at a Wal-Mart in Broken Arrow, Oklahoma, between March and June of 2003, you could have brought home a live RFID chip in the product packaging—and unknowingly starred in a video, too!

P&G is not the only company that's tested spychips on unwitting consumers. Gillette was also caught tagging packages of Mach3 razor blades with some of the 500 million (that's half a billion!) RFID chips it put on order in early 2003. There's also evidence to suggest that other everyday products like Pantene Shampoo, Purina Dog Chow, and Huggies baby wipes may have been tagged with RFID chips and sold to unsuspecting consumers.

Why would anyone want to keep such close track of everyday objects? The answer is simple. Businesses want the technology to give them complete visibility of their products at all times. Having this real-time knowledge would allow them to keep products on store shelves and know precisely what's in their warehouses. They also believe it could help them fight theft and counterfeiting. Theoretically, it could even eliminate the checkout stand, since doorways could scan your purchases automatically when you leave the store and charge them to an RFID-based account.

While some of these goals may sound appealing, the problem is that spychipped products can do a whole lot more, especially once they leave the store with us—and find their way into other areas of our lives.

We've read every pro-RFID argument the industry can make, and we'll be the first to admit the technology could make things more convenient. RFID-enabled refrigerators really *could* keep track of containers of food, warn about expired milk, and generate weekly shopping lists. High-tech washing machines really *could* automatically choose appropriate water temperatures based on instructions encoded in RFID-enabled clothing labels. RFID really *could* help families recover lost pets—and stolen possessions, too.

But when we look at that future, we don't see a twenty-first century Mayberry minus a few entry-level cashiering jobs. The seamy details we've uncovered and will lay out in this book make the spychipped future look more like the ending scene of a gut-wrenching *Outer Limits* episode. The RFID vision that technology companies are selling looks too good to be true—and it is.

Buckle up, readers. We're going to take you on a high-speed, high-tech tour of the past, present, and future of RFID, with plenty of stops along the way at the dirty little secrets *they* don't want you to know.

SPYCHIPS 101

Power to Change the World: It's hard to imagine that a tiny microchip attached to an antenna heralds such enormous change.

—Auto-ID Center promotional brochure,

circa 2002[1]

"THE THING"

New York's Metropolitan Opera house buzzed with anticipation as Leon Theremin took the stage for his sold-out American debut. The distinguished young Russian acknowledged the thunderous applause, then positioned himself behind what appeared to be a wooden podium with four legs, a radio antenna, and a metal loop that jutted from the side. After some tuning adjustments, the physicist-turned-musician gently waved his hands in midair near the antennas of his musical invention, conjuring up haunting wails and moans from an unseen orchestra of radio waves. The ghostly *ooooooooo-weeeeeeee* vibrato sounds were much like the ones that would later become fixtures of 1950s science fiction classics like *It Came from Outer Space* and *The Day the Earth Stood Still.*

The crowds that thronged to Theremin's live performances at the tail end of the Roaring Twenties hadn't had the benefit of watching reruns of B-grade sci-fi movies. So instead of recognizing from the musical cues that something evil or otherworldly was afoot, they did what tragic figures in thrillers often do: They unwittingly welcomed the enemy into their midst.

America's intellectual elite embraced Theremin and even sponsored his ongoing research into radio waves. They never suspected that the Russian-born émigré, nee Lev Sergeivitch Termen, led a double life as a Soviet spy. In addition to marketing his namesake musical invention, the *theremin,* and wooing audiences with his eerie concerts, Lev was secretly relaying intelligence information about the United States' military technology to Stalin in anticipation of a world war. The details of his covert activities and his sudden return to Russia in 1938 are steeped in mystery and speculation.

Whether Lev left New York voluntarily or by force is unknown, but it's likely the KGB was involved. One day, he was carrying out his duplicitous life as usual; the next, he was back in Mother Russia, breaking all ties with his American wife, friends, and benefactors. Some reports indicated that the repatriated Lev later fell out of favor with the Kremlin, was sent to the Gulag, and executed. He was written off as another casualty of Stalin's brutal regime until 1967, when a New York reporter visiting Russia spotted the inventor and sent word home that Theremin was alive and well.[2]

So what was Lev up to all that time? He was developing what some believe to be one of the very first radio frequency identification (RFID) devices. Can you hear the shrill *oooooo-weeeeeee* in the background?

Evidence of his handiwork can be found in a notorious plaque. In the summer of 1945, a group of Russian school children honored U.S. Ambassador Averell Harriman with a beautiful carved wooden replica of the Great Seal of the United States. Harriman's parents must not have passed on the adage "Beware of Greeks bearing gifts" because he proudly displayed the plaque in his embassy residential office, where it hung within earshot of America's Cold War secrets.

The plaque remained in its place of honor until 1952 when the State Department did a precautionary bug sweep of the embassy residence after a redecoration. Nothing was found in an initial pass, but in a secondary sweep, technicians zeroed in on a surveillance device hidden within the plaque. It consisted of an eavesdropping apparatus activated by what was described at the time as a "fantastically advanced bit of applied electronics."[3]

We now know that those "applied electronics" were nothing less than an early form of RFID in its debut performance as a spying technology. Like the spychips raising so much concern today, the Trojan plaque's device was powered by invisible radio waves—in this case, by high-frequency waves beamed at it from a van parked outside the ambassador's residence. Operating the device was as simple as flipping a switch. Because the "Great Seal Bug" lay dormant until stimulated by these invisible waves, it was virtually undetectable. This helps explain why it operated for over six years before being found.[4]

At the time of its discovery, the inner workings of the wireless device were a mystery to American intelligence agencies who were so flummoxed they nicknamed it "The Thing." In a classic case of "spy versus spy," the CIA began its

(PHOTO: COURTESY OF THE NSA MUSEUM)

Great Seal Bug.

own top secret project, codenamed "EASY CHAIR," to learn the secrets of The Thing and unlock its power for themselves.[5]

In 1960, U.S. ambassador to the United Nations Henry Cabot Lodge revealed the true nature of the device, thereby exposing Russia's spying agenda to the world. But it wasn't until years later that the rest of the bizarre picture unfolded. It led back to Leon Theremin, the Russian who had decades earlier thrilled audiences with his musical wizardry. With the advent of Glasnost in the mid-1980s, the formerly repressed Lev revealed details of his years behind the Iron Curtain, including the creation of clandestine spying devices like the Great Seal Bug. We also learned that because of his contributions to covert surveillance, the Soviet Union had honored Lev with a secret First Class Stalin Prize— what was then the Russian equivalent of a Nobel Prize in science.[6]

Though Theremin passed away in 1993 at the ripe old age of ninety-seven, we would like to acknowledge his ingenuity as well. In recognition of his work to develop and promulgate covert radio wave surveillance, we posthumously bestow upon Leon ("Lev") Theremin the title "Father of Spychips," lest the world forget the stealthy legacy of RFID technology.

RFID TODAY

While its past use for surreptitious audio eavesdropping is troubling, RFID's modern incarnation is downright bone-chilling. Radio frequency identification could put us and our information at the mercy of global corporations and government bureaucracies and strip away the last shreds of privacy we have left. Ultimately, its power to reveal, track, and transmit could enslave us.

These are bold statements to make about a technology that's been promoted as merely a cost-efficient way to keep track of items in retail stores like Wal-Mart. Could modern-day RFID really be a wolf in sheep's clothing?

Take a look at how the technology works in this chapter. Then, read in their own words how companies like IBM, Gillette, Intel, and others are planning to fully exploit RFID's potential to track us through retail stores, monitor our use of products in our own homes, and even deliver personalized "spam"-like advertising in places where we can't ignore it.

What Is RFID?

RFID makes it possible to identify and track just about any physical object you can think of—books, car tires, shoes, medicine bottles, clothing, pets, and even human beings.

The "RF" part of RFID stands for "radio frequency" and explains how RFID does its tracking: It uses electromagnetic energy in the form of radio waves to communicate information at a distance. These silent, invisible waves are similar to the radio waves that allow you to listen to your favorite FM radio station. And like other radio waves, they can travel right through windows, wood, and even walls. Of course, radio waves that can go through walls have no problem passing through other items we consider private—like our purses, wallets, backpacks, and clothing.

One industry insider defines RFID as "any device that can be sensed at a distance by radio frequencies with few problems of obstruction or misorientation."[7] This ability to transmit information through solid objects makes RFID far more invasive than the now-familiar bar code with its vertical black and white stripes. Bar code technology uses a laser beam to convey information, so an unobstructed path known as an "optical line of sight" is needed between the bar code and its reader. Since the bar code must be visible to the reader and its pattern must be clean and clear, it's hard to read a bar code someone's carrying without the person knowing about it. In contrast, items equipped with RFID can be located and identified even when other things—like locked doors or sealed envelopes—are in the way.

Tags and Readers

RFID technology can take many forms. It can be incorporated into nails, beads, wires, fibers, or even painted pictures or words.[8] But for the sake of simplicity, let's look at the typical RFID tag that companies want to put on consumer products within the next few years.

There are two main components to an RFID tag. The first component is the tiny silicon computer chip that contains a unique identification number. This

Typical Passive RFID Tag.

(Photo: Katherine Albrecht)

(Photo: Katherine Albrecht)

This tiny vial contains 150 RFID chips manufactured by Alien Technology Corporation. Each tiny chip measures just 0.35 mm square.

chip is often referred to as an "integrated circuit" in scientific circles. The chip can be as small as a speck of dust. In fact, one of the smallest RFID chips in the world is just 0.15 mm square, which is smaller than the period at the end of this sentence.[9]

The second component of the RFID tag is an antenna that's hooked up to the miniature chip. But it's not like the obvious protruding antenna on a transistor radio. An RFID antenna is typically a flat, metallic coil that looks very much like a miniature maze or a tiny racetrack. The antenna coil radiates out from the chip and goes around it in a flat, rectangular configuration, or it may form a long strip, a circle, or an X-shape. The chip and antenna combination, called a "tag" or "transponder," is typically affixed to a plastic surface like an adhesive label or a credit card blank.

Today's RFID tags are generally the size of a thumbnail or larger. The largest ones we have encountered are a little bigger than an index card, and the smallest are about the size of a dime. However, there is a commercially available RFID tag, the Hitachi "mu chip," that measures just 0.4 mm square—antenna and all.[10] This is

(PHOTO: KATHERINE ALBRECHT)

Handheld RFID reader.

half the size of a grain of sand. Though the read range on such tiny tags is quite limited today, as technology progresses, we expect tags will get even smaller and gain new functionality. They are likely to follow the same trend as computers, which have packed more functionality into smaller and smaller footprints over time. To put this in perspective, consider that the computing power that used to take an entire room in an office building can now fit into a hand-held programmable calculator.

Now let's talk about the RFID "reader." Its job is to emit radio waves in a kind of fishing expedition for RFID tags.

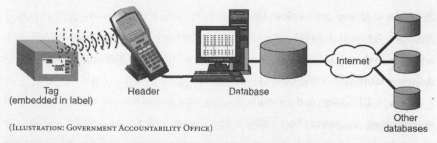

Tag
(embedded in label)   Header   Database

Internet

Other
databases

(ILLUSTRATION: GOVERNMENT ACCOUNTABILITY OFFICE)

Main components of an RFID system include a tag, reader, and database. RFID proponents plan to share data via the Internet.

Here's how a typical tag and reader work together: When an RFID tag gets within range of a reader, the tag's antenna picks up the reader's energy, amplifies it, and directs it to the chip. This energy stimulates the chip to beam back its unique identification number, say 345678 ..., along with whatever other information it was programmed to relay. The reader device captures this information and processes it.

PASSIVE VS. ACTIVE TAGS

What we just described is called a "passive" RFID system, where the tags contain no internal power source of their own. You can think of a passive tag as one that lies around all day doing nothing except waiting for a reader to come along and energize it. A passive tag can't communicate anything unless a reader solicits a signal from it. But don't let the name fool you; since it doesn't need batteries, a passive tag can operate indefinitely, just like Theremin's Great Seal Bug. What's more, you can never tell when someone or something will power it up.

The fact that you only have to power up a single reader device to activate countless little tags makes passive RFID very appealing to RFID engineers, since they can invest in a few readers and buy lots of cheap, disposable tags for the multitude of items they want to track around the planet. Because passive tags are small and lightweight, they can be woven into our clothing labels, sewn into the seams of our undergarments, and even embedded in products inserted into our bodies, like dentures.[11] And that's just the beginning. We'll talk more about these and other disturbing applications later in our saga.

While the passive RFID tag is entirely dependent upon an RFID reader device as a power source, it's also possible to attach a battery to an RFID tag, changing it from "passive" to "active." An active tag, containing its own source of energy, can thereby actively transmit its information payload rather than just lie dormant waiting for a reader. It can also transmit its information further and can transmit more data than the typical passive tag.

Electronic toll collection systems like FasTrack, EZ-Pass, and others use active RFID tags to identify your car as you drive through a tollbooth and automatically charge your account. (We'll discuss this later in Chapter Eleven

▶▶ Tags in Library Books:
What Would Mr. Theremin Think? ◀◀

How ironic. As we were researching the opening for this chapter, Liz purchased a used book about Leon Theremin—Albert Glinsky's *Theremin: Ether Music and Espionage.* To her surprise, she got much more than she bargained for. She found an RFID tag from an Illinois public library on the back flap. The book had been withdrawn from circulation, but someone forgot to remove the tag.

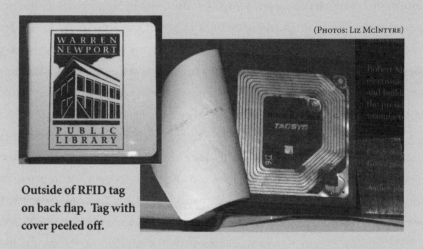

(Photos: Liz McIntyre)

Outside of RFID tag on back flap. Tag with cover peeled off.

Libraries are some of the earliest adopters of RFID, which is troubling because those institutions have helped to preserve the right of all Americans to learn and think freely. Given that library records have already been targeted by overly zealous USA PATRIOT Act Provisions which give FBI agents the ability to seize records of "suspicious" patrons, the library would seem a particularly risky place to open the door to RFID. While some libraries like Warren Newport have adopted RFID without public input, there is a growing chorus of dissent. Privacy advocates like Peter Warfield of the Library Users Association and Lee Tien of the Electronic Frontier Foundation are working to point out not only the privacy threats but also the faulty statistics administrators use to justify spending millions of tax dollars on new library systems.

where we discuss transportation. We'll tell you about how the government uses these toll tags to monitor cars miles from the toll booth, unbeknownst to most drivers.) Keyless remote systems for cars and garage door openers also use active RFID tags.

Typically, having a battery on board an RFID tag makes it bigger, heavier, and more costly, restricting its use to places where bulk and price are not issues. Active RFID tags are popular for use on reusable warehouse pallets and shipping containers, for example, but probably wouldn't work so well on ladies' lingerie. Read range for a passive RFID tag can be anywhere from a couple of centimeters to twenty or thirty feet, depending on the frequency used, the size of the antenna, the amount of power transmitted by the reader, and the environmental conditions. An active tag with a battery can send a signal up to a mile or more. Some very high-powered active tags, like those used to track creatures in the world's oceans, can even transmit information to low orbiting satellites.[12]

As you might expect, passive RFID tags are much cheaper than active tags, so, for the moment, they are the technology of choice for tracking inexpensive items. But engineers are working around the clock to develop inexpensive disposable batteries. The latest breakthrough in battery technology is a flat, printed battery less than a millimeter thick that can be used to power RFID tags.

Precisia, the North American company that is developing the conductive ink for these batteries, characterizes them as "ideal for smart labels, radio frequency identification (RFID) tags and active packaging applications that require an external power source. They serve as an affordable replacement for button batteries in such diverse applications as greetings cards, cereal box giveaways, printed board games, point-of-purchase displays and credit cards."[13] Can you imagine the RFID tag in your cornflake box beaming information down the block?

Spotting Tags

While you can spot many of today's RFID tags if they are sitting in plain sight, RFID tags are easy to hide. Because the tags are usually paper-thin, they can be

▶▶ ARE YOU MY TYPE? ◀◀

Here's some information for the engineers and techies in the crowd. If talk of kilohertz and megahertz makes you dizzy, feel free to skip this box.

RFID readers operate on the same principle you use to tune your radio dial to intercept a particular radio station. For an RFID reader to detect a tag, it must be operating at the same frequency as the tag—and there are a lot of frequencies to choose from. Here are some of the more common RFID frequencies in use today:

Low: 30 to 300 kHz, primarily 125 kHz band and 134.2 kHz. Typically used on animals, including humans, where water content of the body must be taken into account.

High: 13.56 MHz. Mainly used for manufacturing, warehouse and retail store applications.

Ultra High: 300 MHz to 1GHz, primarily 915 MHz. Provides a longer read range, but performs poorly around water and metal. Used in Wal-Mart's warehouses and other "supply chain" applications.

Microwave: above 1 GHz, primarily 2.45 and 5.8 GHz bands. Easily obstructed, works best with line of sight. The 2.45 GHz tags were used in a hospital trial we will discuss later.

Even when their *frequencies* are compatible, tags and readers cannot talk unless they have the right *protocol* or standard. This means they must share the same language in order to understand each other.

---

sandwiched between the layers of cardboard in boxes so they are visually undetectable. Manufacturers can embed the tags into everyday items during manufacturing by heat-sealing them into plastic, incorporating them into rubber, or even embedding them in a shirt label. Amazingly, the industry has even invented flexible RFID tags for clothing. Fine metallic threads sewn into the fabric act as the antenna.[14]

RFID tags could appear on the surface of products, yet still be visually undetectable. While most RFID antennas use a metallic coil or strip, a company called Flint Ink has developed a spray-on conductive ink that can serve as an RFID antenna. They can put the tiny chip on top of this gray matte-looking surface, then cover it with regular packaging ink so the consumer would never see it.[15]

Even the RFID chip itself could be printed one day. Scientists have discovered conductive organic polymers that can be dissolved in ink. Siemens, the global technology giant, hopes this will open the door to printed electronic circuits that will replace today's silicon circuits. Not only would these spychips be far less expensive, they would be flexible. According to Siemens, a futuristic chip "woven into a sweater could, for example, inform the washing machine of the water temperature it needs to provide."[16]

Tags can even be "chipless." Currently, they don't perform as well as their chipped counterparts, but time and research are narrowing that divide. Even without a chip, these tags can still be read through a brick wall.[17]

If you're getting the sense that RFID tags could be stealthy and hard to spot, you're correct. That's why CASPIAN developed sample consumer labeling legislation, the *RFID Right to Know Act*, that requires disclosure if a product contains an RFID tag. We believe consumers have a right to know if the products they interact with and buy could relay information about them or their shopping habits without their knowledge or consent. We will discuss this legislation in more depth in Chapter Seventeen.

Spotting Readers

Just as tags can be well hidden, so can the reader devices that interrogate them. In fact, readers can be even harder to find because they don't have to fit in small packages or conform to various product designs. Since optical line of sight isn't needed for radio wave transmission, RFID readers can be embedded in doorways, woven into carpeting and floor mats, hidden under floor tiles, embedded in ceiling tiles, incorporated into shelves, and placed behind store displays.

We've already gotten a taste of what's in store. The RFID "smart shelf" reader device made infamous by Gillette (we'll cover this photo-snapping spy shelf later in the book) looks a lot like a medium-sized pager. While early versions

▸▸ NOT EVERYTHING THAT LOOKS LIKE
AN RFID TAG IS ONE ◂◂

Commonly used anti-theft devices resemble RFID tags but lack the computer chip that can store information (at least for now). More on this in Chapter Four.

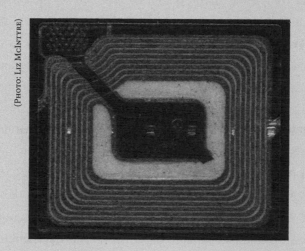

(PHOTO: LIZ McINTYRE)

Typical RF EAS (anti-theft) tag. This is not an RFID tag.

(PHOTO: LIZ McINTYRE)

Typical acousto-magnetic EAS tag found in DVDs—not an RFID tag.

had rigged the reader under existing shelving, leaving evidence of wiring exposed, the newer versions incorporate the reader directly into the shelf itself, as part of the shelf design. More conventional RFID readers designed for use in retail stores and supermarkets can look very much like the current equipment that reads bar codes, so a transition to RFID might not be apparent at the cash register. In fact, the handheld bar code price verifiers that retail employees carry around while completing inventory and the bar code scanners (handheld and stationary) at the point of sale could read both bar codes and RFID tags if upgraded. Outside the store, the location of reader devices would only be limited by the imagination. Why not incorporate readers in vending machines or benches just outside the store? Readers could be blended into the landscape, perhaps hidden in artificial landscape boulders or signs.

There is already a bevy of handheld, portable RFID reader devices available that are not hidden but look very much like other wireless devices. There are also readers that are made to be incorporated into handheld PDAs like Palm Pilots and cell phones like the Nokia 5140.

RFID readers may even invade our homes if proponents have their way—presumably with the knowledge and consent of consumers. There are prototype appliances with RFID readers on board in anticipation of the day when every product manufactured on earth will have an RFID tag. For example, in its "Home of the Future," Microsoft demonstrates a microwave oven that talks with spychipped frozen entrees to assure correct power settings and cook times. The reader in the family refrigerator keeps track of its contents.

Before you sign up for one of these whiz-bang futuristic appliances, you need to understand the implications. Coming up in the next chapter, we'll be laying out the industry plan for all this technology. We'll forewarn you that it isn't pretty!

3

THE MASTER PLAN

The Auto-ID Center has a clear vision—to create a world where every object—from jumbo jets to sewing needles—is linked to the internet. Compelling as this vision is, it is only achievable if the center's system is adopted by everyone everywhere. Success will be nothing less than global adoption.

—Helen Duce,

an Auto-ID Center associate director[1]

The ability to surreptitiously collect a variety of data all related to the same person; track individuals as they walk in public places (airports, train stations, stores); enhance profiles through the monitoring of consumer behaviour in stores; read the details of clothes and accessories worn and medicines carried by customers are all examples of uses of RFID technology that give rise to privacy concerns.

—EU working document on RFID, January 2005[2]

SHADES OF THE FUTURE

You can blame it on lipstick. Specifically, Oil of Olay's ColorMoist Hazelnut No. 650, a Procter & Gamble product. Kevin Ashton was a young brand manager in charge of launching the new shade back in 1997, and he couldn't keep it in stock. It was far too popular. But while it was selling out on the store shelves, there was plenty of it ready to ship back at the warehouse. What to do?

Ashton searched high and low for ways to solve his supply chain problem. A year later, he learned about a technology called RFID that was already being used for toll collection and building access. It occurred to him that this technology might help solve his problem.

He sought out the advice of two researchers from MIT who had been toying with ways to miniaturize RFID technology, Professor Sanjay Sarma and Dr. David Brock. The three huddled over the lipstick problem and emerged with the idea of putting a computer chip bearing a unique identification number on each tube of lipstick. Tracking each specific tube of lipstick would allow them to keep better tabs on the inventory than would be possible with a bar code that merely identified types of products.

Computer chips were fairly large and expensive at the time, so there was probably plenty of laughter over the idea at Procter & Gamble's Cincinnati headquarters—at least at first. But the plan soon won favor in the boardroom. With funding from Procter & Gamble, Gillette, and the Uniform Code Council (the bar code people), the trio founded the MIT Auto-ID Center in October 1999, and Kevin Ashton took the helm as its director.

With corporate heavyweights on board, the center's plan quickly expanded beyond keeping the world's women in hazelnut lipstick. They realized that this powerful new technology could make it possible to track *everything*. They had a vision:

> The Auto-ID Center's vision is a world in which low-cost RFID tags are put on every manufactured item and tracked using a single, global network as they move from one company to another and one country to another. Indeed, we envision individual items—cans of Coke, pairs of jeans and car tires—being tracked from the moment they are made until the time they are recycled. This will give manufacturers and retailers near-perfect supply chain visibility. It will eliminate human error from data collection and enable companies to reduce inventory, make sure product is always on the store shelves, and reduce lost, stolen or misdirected goods. It will open a new

world of convenience for consumers, who one day may be able to check themselves out at a supermarket in seconds. In short, it will transform the way we do business and the way we live.[3]

Of course, implementing this vision would require an infrastructure for tracking everyday objects. Not only would RFID tags have to be affixed to everything, but RFID tag readers would have to be everywhere—in factories, in warehouses, on trucks, in storerooms, in retail spaces, in homes, and even in garbage trucks. To make the information actionable, the Auto-ID Center would also need to develop a way for those RFID tag readers to communicate tag information in real time, all the time, to those managing the supply chain.

Thus was born the "Internet of Things."

This proposed "Internet of Things" wasn't to be a new network; it was to be built on top of the existing Internet. But what was earth-shatteringly new was the revolutionary idea that *inanimate objects* would be endowed with the ability to talk to manufacturers, retailers, and even each other.

A Number for Everything

So what would an object like a shoe, a shirt, a box of cereal, or a can of Coke have to say? Quite a bit. It's all about numbers.

The plan called for each item's computer chip to contain a unique number, known as an EPC or "Electronic Product Code." (In contrast, our present bar code numbering system is called the UPC or "Universal Product Code.")

Like the bar code, this new system would contain information about the manufacturer (say, Coca-Cola) and the product (say, twelve-ounce can) but

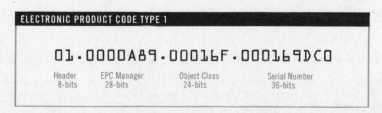

SOURCE: DAVID BROCK, "WHITE PAPER: THE ELECTRONIC PRODUCT CODE," AUTOID CENTER, 2002 P.6.

with a crucial new twist: It would also contain a unique serial number not shared by any of the other trillion items on the planet—not even its fellow cans of Coke.

So enamored was the Auto-ID Center with the idea of uniquely identifying all objects, they developed a system that could number every item produced on earth for the next thousand years—each with its own unique ID number and no repeats.

To see what a monstrous undertaking this is, consider the size of the world and the overwhelmingly vast number of objects in it. Then, consider what a daunting challenge it would be to uniquely number them all. While many numbering schemes were considered, the EPC developers finally settled on a ninety-six-bit code. (That's a string of ninety-six zeroes and ones, or, said a different way, two to the ninety-sixth power.) The developers tell us this code is adequate to uniquely number a mind-blowing "80 thousand trillion trillion objects—more than sufficient for man-made physical products."[4]

Note from their chart below that the Auto-ID Center made provisions for numbering not only grains of rice and razorblades but also every human being on the planet, too. (We've devoted a whole chapter to people-tracking later in the book.)

BITS	UNIQUE NUMBER	OBJECTS
23	6.0×10^6 per annum	Automobiles
29	5.6×10^8 in use	Computers
33	6.0×10^9 total	Humans
34	2.0×10^{10} per annum	Razor blades
54	1.3×10^{16} per annum	Grains of rice

This chart shows that thirty-three bits are all it would take to assign a unique number to six billion (6.0×10^9) human beings. (SOURCE: AUTO-ID CENTER WHITE PAPER: "THE ELECTRONIC PRODUCT CODE (EPC): A NAMING SCHEME FOR PHYSICAL OBJECTS" BY DAVID BROCK[5])

Of course, it's possible for a computer chip to store a lot more than just a number. But since the Auto-ID Center's goal was for companies to put one of these tags on every item in their inventory, the price had to be low—low enough that no one would think twice about using them to track even packs of gum and sewing needles. When they said everything, they meant *everything*.

WHAT'S IN A NUMBER?

Kevin Ashton once characterized the low-priced chip envisioned by the Auto-ID Center as "the amoeba of the wireless computing world"[6]—and in a strange way, he was right. Like an amoeba that can wreak havoc far out of proportion to its size, the simple number on an RFID chip is far more powerful than it appears to the casual observer.

Assigning a unique serial number to everyday objects is like giving them Social Security numbers. It makes it possible for businesses to create a unique data file for each item that can store virtually unlimited amounts of information about it. Or, as MIT Media Lab Researcher Joseph Kaye described it, it's like giving every can of beans its own webpage:

> The next step will come when you purchase a product which has its own individual webpage. A can of beans will come with its own individual webpage detailing such information as production date, transport history, and time spent on the shelf, all entered automatically as it moves along the retail chain. Two apparently identical packets of rice you purchased on two trips to the supermarket can have entirely different histories of transport, storage, and origin.[7]

Since each can of beans would be connected to the "Internet of Things," through its RFID tags, it literally *could* have its own webpage. That is the goal, and it's very real. Verisign, the company that manages webpage addressing for the entire Internet, has already agreed to oversee addressing for the Internet of Things.[8]

Glance around the room you're in and pick a random object. What might a webpage connected to the Internet of Things for that item look like? We'll offer up Liz's scarf as an example. Fortunately, it does not have an RFID tag in it. But if it did, the serial number on the tag might link to a data file (or a series of data files) set up by the manufacturer containing the scarf's production and shipment history, style, fabric content, and washing instructions. If the retailer who sold the scarf participated in the Internet of Things, it might record the sales receipt information noting the date, place, and price at which it was sold, and perhaps even the name and credit card number of the person who bought it.

Whenever Liz wore the scarf and came within range of a reader device connected to the Internet of Things, its chip could communicate directly to its Internet data file. It could check in to update its status and log information about its surroundings and activities. Updates could include the date and time it was seen, the location of the reader, other clothing items detected nearby, purchases Liz might have been making at the time, and who was standing near her when the scan occurred.

Thanks to advances in computer technology and bargain-basement prices of data storage, there would be virtually no limit to the amount of information that could be stored in this way. If this seems hard to believe, consider that by 2004 Wal-Mart's database already contained twice as much data as the Internet.[9]

The scarf example illustrates how RFID tags make the bar code look downright primitive by comparison. This extraordinary potential was not lost on Ashton's group. The light bulb went on: Imagine what would be possible if these tags became as widespread as bar codes. Heck, why not replace the bar code altogether? So the Auto-ID Center nicknamed its RFID initiative "The Journey to Discover What Will Follow the Bar Code."[10]

But It Isn't a Bar Code!

The problem with calling RFID an "improved bar code" is that it's not true. RFID differs from bar codes in three important ways:

Unique ID. With bar codes, every twelve-ounce can of Coke has the same UPC number (i.e., bar code), but proposed RFID tags would assign each individual can of Coke its own *unique* serial number. These serial numbers could be captured at the point of sale and recorded with the identity of the

---

▶▶ Who Wants to Know? ◀◀

Who would want to scan someone else's RFID tags? While big business will tell you nobody plans to read RFID tags after goods are sold, we don't believe it for a minute. We've identified three groups with a big interest in secretly snagging spychip data from the things you own: marketers, government agents, and criminals.

Marketers want tag data to identify you and profile your possessions so they can target you with marketing and advertising material wherever you go. *Government agents* crave the power of hidden spychips to monitor citizens' political activities and whereabouts. And, of course, *criminals* can't wait to identify easy marks and high-ticket items by scanning the contents of shopping bags and suitcases at a distance.

---

purchaser as gleaned from a credit card or frequent shopper card. Such linkage could lead to global item registration where the ownership trail of virtually every item on earth could be recorded in a database and used to monitor peoples' travels and activities.

Remotely readable. RFID tags can be read from a distance by anyone with the right reader device, right through people's clothes, wallets, backpacks, or purses—without their knowledge or consent. It creates a form of x-ray vision that could enable strangers to identify people and the things they're wearing and carrying. In the spychipped future, readers could be hidden in stores, public buildings, homes, and even outdoor spaces like parks to electronically frisk you as you pass by, taking notes on everything in your possession—right down to the size and color of your underwear. We'll talk about doorway portals and other devices that could make this possible in an upcoming chapter. Since the reading would be silent and invisible, you would never be the wiser.

Health risks. Unlike the optical readers associated with bar codes, RFID readers emit electromagnetic energy over wide swaths. Since global corporations hope to embed RFID readers into walls, floors, doorways, shelving— even in the refrigerators and medicine cabinets of our homes—we and our children would be continually bombarded with the energy emanating from these devices. Medical researchers have begun to raise questions about the long-term health effects of this type of chronic exposure to low levels of electromagnetic radiation.[11]

Buying "Intimate Access"

The societal downsides of making everything trackable and possibly jeopardizing human health did not deter the Auto-ID Center. On the contrary, once they realized the power they could unleash, they wanted it even more. So they sought the support and buy-in of the world's largest manufacturers and retailers to make their vision of ubiquitous item identification a reality. Since only volume purchases would create the economies of scale needed to drive down the price of chips, the Auto-ID Center pounded the pavement to find additional sponsors willing to cough up the $300,000 for "intimate access to groundbreaking research."[12]

And succeed they did! What started as a farfetched pipe dream in 1999 had mushroomed into a corporate juggernaut by 2002, with sponsors like Wal-Mart, International Paper, Home Depot, Intel, Pepsi, Coca-Cola, Target, Tesco, Phillip Morris, Unilever, Kodak, and UPS on board. The Center even boasted the United States Postal Service and the Department of Defense as sponsors. By this time, the Auto-ID Center's research was deeply advanced, and live trials in stores were already underway, though very few people besides the sponsors themselves knew anything about it. In fact, they were counting on keeping their trials under wraps, since it wasn't in their best interest to alert the public to the privacy-robbing technology they were quietly preparing to unleash.

Fortunately for consumers, it was around this time that Katherine stumbled across an article that revealed a startling glimpse of their enormous plans. In the article, a senior vice president of the market research firm ACNielsen (the people who monitor television viewing habits) boasted, "Where once we collected purchase information, now we can correlate multiple points of consumer product purchase with consumption specifics such as the *how*, *when* and *who* of product use."[13]

Wait a minute. The "*when* and *who* of product use"? That sounded suspiciously like monitoring people's use of products in their homes. How could they possibly do this?

Katherine discovered that the heart of this project was based at MIT, just down the street from Harvard where she was working on her doctorate. So she slipped into a few of their strategy sessions, including the November 14, 2002, Auto-ID Center Board of Overseer's meeting. As the event unfolded, Katherine could barely believe what she heard.

Dick Cantwell, the Auto-ID Center chairman and Gillette's vice president of global business management, stood up and told the crowd of executives he had some big news. After years of preparation, RFID had finally advanced beyond the laboratory phase and was ready to be released into the world. Gillette had put in a purchase order for half a *billion* RFID chips and was ready to begin using them on its products. The packed crowd fell completely silent as everyone processed the news. It was the stunning moment they had all been waiting for. Millions of dollars of investment and three years of development had come to fruition. Gillette was about to release five hundred million spychips into the real world.

The next bombshell was the Auto-ID Center's announcement that it would be passing control of the RFID initiative to the Uniform Code Council. If you recall, that's the organization that manages the bar code. An era was coming to a close. Spychips had outgrown their period of incubation at MIT and were now graduating to the big boys with the power to put them on everything. The EPC network was poised to become the new global standard for item identification, but average consumers knew nothing about it. How would they respond when they found out?

Helen Duce, one of the Auto-ID Center's associate directors, reported on her consumer research. She told the crowd she had some good news and some bad news. The bad news was that the people they had interviewed around the world weren't going to like RFID. What's more, they were distrustful that global corporations and governments would use it responsibly. The good news, she said brightly, was that people felt hopeless to fight against the technology. She reassured the ruffled executives in the audience that consumers would be unlikely to fight RFID—unless, of course, some privacy activist came along to stir them up.

As if all of this weren't enough, then an executive from Intel stood up at that point and offered the following advice about consumer advocates: "We should bring them in so that they . . . I don't want to use the word 'co-opt' but . . . we should make sure we deal with their [issues]."[14]

Horrified, Katherine sat in the audience and took careful notes. Someone had to alert the public. Clearly, businesses had plans they wanted to pursue at all cost—plans they realized consumers would fight if they knew the details.

Patents Tell the Story

When word began to get out about RFID, industry players tried to lie about their Machiavellian goals for the technology, saying they would never use it to

---

▶▶   We Swear It Works   ◀◀

I hereby declare that all statements made herein of my own knowledge are true and that all statements made on information and belief are believed to be true; and further that these statements were made with the knowledge that willful false statements and the like so made are punishable by fine or imprisonment, or both. . . .[15]

So reads the official U.S. patent application form.

The patent process is a serious legal undertaking. Inventors filing an application must take an oath swearing that the information they present is truthful and that the invention meets the requirements to be patentable. To qualify for a patent, an invention must be shown to be both "useful" and "operative." In other words, it has to work. Any machine or process which cannot perform its intended purpose cannot be called useful, and therefore should not be granted a patent.[16]

While not all patent filings are approved, and many never pan out financially for their inventors, they can still provide rich insights into how a company operates, its ethical standards, and its long-term thinking, goals, and priorities.

track people. But Liz found the noose the industry had put around its neck: blueprints detailing precisely the methods global corporations would use to make our worst nightmares come true. Big companies like IBM, Procter & Gamble, NCR, and other visionaries have all filed patent documents that provide scandolous snapshots of how they're proposing to spy on consumers with this technology.

If there's a corporation that should understand spychips and their potential, it's IBM. With RFID labs established in countries around the globe, the technology giant is one of the most heavily invested companies in RFID projects. So when it files for a patent involving the technology, we pay attention.

On May 3, 2001, IBM inventors filed patent application #20020165758, IDENTIFICATION AND TRACKING OF PERSONS USING RFID-TAGGED ITEMS.[17] That patent application spells out one of the key problems with RFID: It can all too easily be used to track people. How? IBM details a way to collect RFID numbers at the cash register and store the numbers in a database. Then, later on, "the exact identity of the person" can be determined from the tags and "used to monitor the movements of the person through the store or other areas." IBM explains:

> Previous purchase records for each person who shops at a retail store are collected by [cash register] terminals and stored in a transaction database. When a person carrying or wearing items having RFID tags enters the store or other designated area, a RFID tag scanner located therein scans the RFID tags on that person and reads the RFID tag information. The RFID tag information collected from the person is correlated with transaction records stored in the transaction database according to known correlation algorithms. Based on the results of the correlation, the exact identity of the person or certain characteristics about the person can be determined. This information is used to monitor the movement of the person through the store or other areas.

The patent application goes on to describe how RFID tags could be used to identify a person's age, race, gender, and income bracket:

. . . instead of determining the exact identity of the person, some characteristics such as demographics (e.g., age, race, sex, etc.) about the person may be determined based on certain predetermined statistical information. For example, if items that are carried on the person are highly expensive name brands, e.g., Rolex watch, then the person may be classified in the upper-middle class income bracket. In another example, if the items that are carried on the person are "female" items typically associated with women, e.g., a purse, scarf, pantyhose, then the gender of the person can be determined as a female.

Once IBM has the information, they could even use it to see if you're from out of town or how long you've owned that pair of underwear:

When the system is configured to identify the general demographical information about the person, information such as the gender, age, social economic status, geographic location where they probably purchased the products, how long the products have been in service, etc., may be determined.

IBM has even developed a device named the *"person tracking unit"* that can zero in on your RFID tagged products and use them to watch you like a hawk:

Once the exact identity or some demographics or other characteristics of the person have been determined, the person tracking unit relies on this information to track the person as the person moves through the roaming areas. The person tracking unit may assign a tracking number to each identified person and store the tracking number in association with the collection of RFID tagged product information.

So why would IBM suggest obtaining all this information? Well, here's one reason:

Once the movement of the person can be monitored based on the RFID tags carried on the person, the tracking information can be used in a number of different ways. For example, it can be used to provide targeted advertising to the person as the person roams. . . .

That's enough to steam even the most tolerant consumer. But it gets worse than just advertising. IBM inventors suggest that the government could track suspicious persons in public places via the RFID tags in the things they are wearing and carrying. This is particularly shocking in light of the fact that IBM leased cutting-edge punch card technology to the Third Reich so that Hitler could track Jews and their property.[18] Of course, if the government can track people deemed suspicious, it theoretically could track everyone else as well:

Although the systems . . . of the present invention . . . have been described in context of a retail store, it can be applied to other locations having roaming areas, such as shopping malls, airports, train stations, bus stations, elevators, trains, airplanes, restrooms, sports arenas, libraries, theaters, museums, etc.

Yikes! Has IBM really been working on ways to track us in elevators and libraries?

Yes, and much more. And it's all in the public record.*

Unfortunately, IBM is just one of many companies with such big ideas. We will share equally astounding patent documents from other well-known companies in the chapters that follow to prove to you that there's an agenda out there—and apparently all of humanity is slated for the receiving end of it.

In the coming chapters, we have much more proof of what the RFID industry is planning and why it's crucial to fight back.

---

* Perhaps more than any other document, this IBM patent application spells out the nightmare vision of RFID. We believe it is so important that we will revisit it again later in this book, and we reprint it in its entirety on our website at Spychips.com.

4

THE SPY IN YOUR SHOE

[T]he widespread use of RFID tags on merchandise such as clothing would make it possible for the locations of people, animals, and objects to be tracked on a global scale—a privacy invasion of Orwellian proportions.

—IBM U.S. patent application 20020116274[1]

SPYCHIPPED CLOTHING LABELS

RFID trade shows are the go-to events for spychippers and business executives thinking of joining their ranks. There, RFID industry players market their wares, brag about their latest developments, and give hands-on access to their technology. Naturally, that's where we do some of our best research.

In 2004, we traveled to Chicago's Navy Pier conference center to scope out one of the biggest of the annual shows, Frontline Solutions. We expected to find plenty of insights and surprises, and we were not disappointed.

The shocker at that show was a brazen display of prototype Checkpoint RFID-laced clothing labels and "hang tags" bearing the names of companies like Calvin Klein, Champion, Carters, and Abercrombie & Fitch. (Checkpoint is the manufacturer of anti-theft tag systems and now RFID systems, too, as we'll

This Calvin Klein fabric label was among the
items featured at the Checkpoint RFID booth.
Viewed from the front, it looks like an ordinary
clothing label that would be sewn into the collar
of a shirt or sweater.

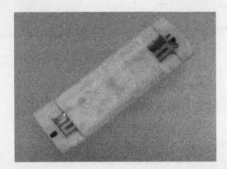

Viewed from the back, it is apparent
that the Calvin Klein clothing label
contains a hidden RFID device. (Note
the metallic antennae extending from
either side.)

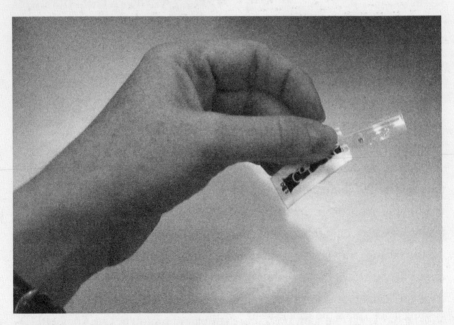

When the Calvin Klein clothing label is opened, the RFID device it contains can be
clearly seen. Note the computer chip in the center of the tag and the metallic
antenna extending from it. This chip contains a unique ID number that can be
read remotely.

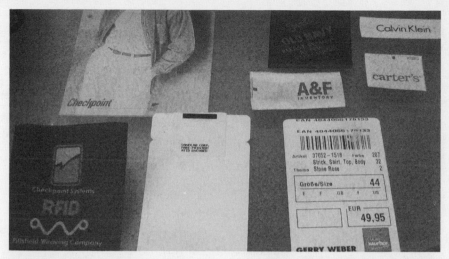

At the Frontline Expo 2004 conference, Checkpoint revealed that one of their major clothing clients was secretly working on plans to incorporate item-level RFID tags into all their merchandise. Could it be Abercrombie & Fitch? Old Navy? Calvin Klein? Carter's? Champion? Apparently the brand doesn't want its RFID involvement publicized; Checkpoint's lips were sealed.

This label, reading "Checkpoint Systems RFID," was sewn into a Champion athletic wear jacket on display at the Checkpoint exhibit. Champion is owned by Sara Lee Corporation, one of the earliest companies to invest in the development of RFID technology.

Interior view of a Checkpoint RFID clothing label. This is the same type of label that was sewn into the Champion jacket pictured above. While it doesn't appear clearly here, the tag contains woven circuitry inside.

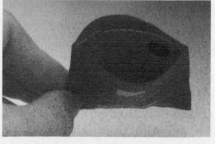

(Photos: Katherine Albrecht)

explain later.) As you'll see in the photos below, the fabric labels looked just like the ones you might find in your own closet, neatly sewn into the neck of a shirt, a jacket, or a baby's romper. The cardboard hang tags looked just like any others you might find dangling alongside the price tags on clothing at any store in the world. The difference was that these tags were souped up, spychipped, and remotely readable.

These were more than mere prototypes. The Checkpoint representative boasted that a well-known client planned to put Checkpoint RFID tags on all its clothing items. He wouldn't reveal the name of the company, but he dropped a few clues.

Abercrombie & Fitch?

The Checkpoint representative referred to the company as a "well-known national retailer." Other clues included two dark blue prototype hang tags—the only two tags in the display where there was any apparent attempt to conceal the company name. Though, in retrospect, the shoddy taping job that left only the Abercrombie & Fitch collegiate logo at the bottom could have been Checkpoint's way of adding drama to the presentation.

When Liz held up the poorly obscured tags and asked if the "mystery" company planning to tag everything was Abercrombie & Fitch, the representative couldn't hold back a telling smile. While he claimed his lips were sealed, his manner suggested that we'd guessed the answer.

Abercrombie & Fitch would be just the company to do it, too. These are the same guys who have peddled "soft porn" to kids, glorified teen nudity, and portrayed group sex parties in their advertising campaigns. Their bad behavior has sparked boycotts from Christian organizations, Asian-American groups, and even the state of West Virginia. In addition to using sexually explicit catalogs to market their clothing, they have printed up culturally insensitive shirts with slogans like "It's All Relative in West Virginia," and "Wong Brothers Laundry Service: Two Wongs Can Make It White."

Whether or not you find their products offensive, we can probably all agree that they've been willing to ignore consumer opinions in the past. Spychipping clothes, a key concern in consumer studies, would be completely *in* character for this organization.

Turns out our suspicions were justified. A few months after the conference, Abercrombie & Fitch let the cat out of the bag. In an article titled "U.S. Clothes Firm Comes Clean on RFID Plans," A&F admitted to trialing RFID technology. "We're testing it," a company executive told Silicon.com reporter Jo Best. "We've got a couple of plans, we're looking carefully at it. . . . Everybody is."[2]

He's right. Major clothing manufacturers like Calvin Klein, Carter's, and Levi-Strauss, along with retailers like Wal-Mart, Target, and Tesco, are eager to tag every shirt, every pair of pants, and every shoe they sell—not that they want people to know about it yet.

Getting the Word Out

As soon as we returned from the trade show, we issued a press release and posted the shocking clothing label photos online, generating a rash of news stories. The industry was none too pleased. Advanstar Communications, the company that organized the Frontline Solutions show, sent an e-mail demanding that we...

> ... remove all unauthorized photos that you obtained at Frontline Solutions Conference & Expo from your websites, www.spychips.com, www.spychips.org, www.nocards.com, www.nocards.org and any other websites under your management and that you refrain from making the photos available to anyone else. If these photos are not promptly removed from your websites, then Advanstar will not allow you access to Frontline or any RFID-related Advanstar exhibitions or conferences in the future.

Despite Abercrombie's later admission, it was obvious that the RFID industry didn't want the rest of the world to see evidence of their secret tagging plans. But consumers have a right to know what's in store, so we stood fast. Rather than bow to censorship demands, we not only kept the photos on the site, but we also issued a press release indicating that we would be posting additional photos from the show, including ones documenting the item-level tagging of Huggies baby wipes, Kimberly-Clark diapers, Nyquil cold medicine, CVS vitamins, Similac baby formula, and Lanacane cream.

Clearly, the industry is concerned about a backlash—a well-founded fear considering the fiasco that forced Benetton into retreat in 2003.

THE UNITED COLORS OF BENETTON

Benetton is the Italian clothing manufacturer that sells its pricey clothing in over one hundred countries around the world, marketing its brand with the distinctive "United Colors of Benetton" slogan in its glossy advertising. Of course, it's also infamous for its "United Killers of Benetton" campaign, in which it used death row inmates to market its colorful sweaters a few years back. But that was just the start of the company's troubles.

When Katherine got wind that Benetton was planning to put spychips in its Sisley line of women's clothing and barely-there undergarments back in the spring of 2003, she led CASPIAN on the warpath against the company, calling for a worldwide boycott of their clothes.

CASPIAN's "I'd Rather Go Naked" campaign was the first shot fired across the bow of the RFID privacy wars. Within weeks, the story had been covered by National Public Radio, *Business Week*, Reuters, and dozens of media outlets across the country. But the boycott didn't stop there. It was picked up by the *Hindustan Times*, *The Scotsman*, and featured in the Netherlands, France, Germany, and as far away as Tasmania.

Reeling from a worldwide backlash, Benetton was forced to cancel its plans to spychip over *fifteen million* garments and reassure shoppers that its clothing would be spychip-free. The victory was cheered as a win for consumers everywhere. We had put a temporary stop to the industry's invasive spychipping plans.

This might explain why Frontline was so anxious to suppress the evidence of a renewed effort to tag clothing in 2004. The Benetton boycott had bought us some time, but the Checkpoint labels made it clear that companies were once again playing with fire—this time with a bit more stealth.

The Spychip Agenda: Tag Everything!

Unless consumers put a stop to it, product manufacturers and retailers will put spychips into everything manufactured on earth. If and when this happens, your possessions could be used to identify you, profile you, and track your movements. But that hasn't happened yet.

Or has it?

Because RFID tags can communicate right through solid objects, and can be easily hidden, they could already be planted in the things we wear and carry today. We have seen prototype RFID tags concealed between layers of cardboard, incorporated into product labels, and encapsulated in plastic. And, as you've now seen with your own eyes, there are prototypes of clothing labels with RFID tags hidden inside.

This is more than just a theoretical concern. There are at least two confirmed cases in which items on shelves in retail stores contained hidden RFID tags that were used to spy on people, and we have hints that other consumer products have also been secretly tagged.

First, let's look at the confirmed cases.

Gillette's Photo-Snapping Smart Shelf

In July 2003, we helped the UK *Guardian* newspaper expose a scandal involving Tesco, the UK's largest retail store, and Gillette, the razor blade company.[3] Gillette was quick to put its large order of spychips to use, hiding them in the glued cardboard packing of Mach3 razor blades. An RFID-enabled Gillette "smart shelf" could sense when consumers picked up the spychipped packages. SNAP! A camera relayed a close-up mug shot of each shopper's face to store security personnel. (After all, anyone could be a shoplifter, they figured, so why not watch 'em all?)

A second camera snapped the shopper's photo at checkout. If someone was seen picking up razor blades from the shelf but wasn't seen paying for them, that person was flagged as a potential shoplifter. (Pity the customer who abandoned the pricey blades in a magazine rack at the checkout or who had a spouse pay the bill.)

All this photographing and tracking was done secretly, without the knowledge of customers shopping at a Tesco store in Cambridge, England. But all it took was a little exposure. The companies' "guilty-until-proven" innocent approach to crime prevention didn't go over too well with consumers, who protested outside the store after the *Guardian* broke the story. Like cockroaches, the chips and shelves disappeared as soon as the light was shined on them.

But, of course, while the shelves may have disappeared, Gillette and Tesco are still big RFID backers, promoting the concept behind the scenes. Gillette's Dick Cantwell publicly talks up the technology at every opportunity, and Tesco, the world's third-largest retailer with thousands of stores across Europe and Asia, has since gone on an RFID "shopping spree," buying 20,000 readers and antennas to install in 1,300 of its stores.[4] Since both companies have been unrepentant for their past deeds and have failed to address consumer concerns about their use of spychips, CASPIAN has launched worldwide boycotts against them as well. Details are available at BoycottGillette.com and BoycottTesco.com.

Photo-snapping spy shelves were also slated for the United States. Katherine took pictures of one in a Brockton, Massachusetts, Wal-Mart store. That shelf disappeared overnight, too, after Elaine Allegrini of the *Brockton Enterprise* called company officials for an explanation and ran a front-page story. "Shelf? What shelf?" they responded.[5]

Hiawatha Bray of the *Boston Globe* doggedly pursued an acknowledgement of the shelf from Gillette and Wal-Mart and confronted them with Katherine's photos. They suddenly remembered the shelf but denied they ever plugged it in. (We bet they never inhaled, either.)

PROCTER & GAMBLE TESTS ITS TECHNOLOGY ON UNWITTING SHOPPERS

In November 2003, *Chicago Sun-Times* reporter Howard Wolinsky broke the shocking story of how Procter & Gamble conducted a secret RFID trial at a Broken Arrow, Oklahoma, Wal-Mart store.[6] They'd taken Kevin Ashton's lipstick tracking plans and put them into action. Customers who bought Lipfinity lipstick from that store between March and mid-July of 2003 got something extra in the bargain: a live RFID tag hidden inside the product packaging.

A Gillette razor package prototype with RFID tag concealed inside.

Close-up of tag concealed in Gillette package.

(PHOTOS: LIZ MCINTYRE)

(PHOTO: KATHERINE ALBRECHT)

Procter & Gamble Lipfinity lipstick package and an adhesive RFID tag similar to the ones that triggered a webcam in a Broken Arrow, Oklahoma, Wal-Mart store.

A video webcam was trained on female shoppers to watch them as they picked up the spychipped lipsticks. The live feed of the women's images was beamed directly to the offices of Procter & Gamble executives 750 miles away. ("Please close my door and hold my calls, Ms. Jones.") It was a perfect illustration of how easy it is to set up a secret RFID infrastructure and use it to spy on people. Had an informant not blown the whistle, they would probably have gotten away with it. Did they apologize? Hardly.

When Wolinsky, the investigative reporter, contacted Procter & Gamble, they flatly denied the trial had ever taken place. But P&G was no match for Wolinsky's sleuthing skills. He uncovered photos and other proof that reduced the company's public relations department to incoherent mutterings. When denials no longer worked, they said customers shouldn't object to such a trial since Wal-Mart's standard store placards stated that the premises were under surveillance.[7]

Under surveillance? What an understatement!

In addition to these documented cases, we have run across photos and written evidence suggesting that other products have contained hidden spychips, too. These include Huggies baby wipes, Pantene shampoo, Caress soap, Right Guard deodorant, Wisk laundry detergent, and bags of Purina dog food.[8]

RFID SLIPPED UNDER OUR NOSES?

Let's say these companies succeeded in spychipping our shirts and underwear—what would it take to create a worldwide network of devices to read all these tags? Frighteningly little. The next time you walk through the doors of a major retail store like Wal-Mart or Target, take a look to the left and to the right. You'll see anti-theft security portals standing guard. What few people realize is that these gates could turn into RFID readers overnight.

Stores have used security portals for a number of years to detect swiped goods. Two companies, Sensormatic, a subsidiary of Tyco, and Checkpoint Systems, the company we caught displaying spychipped clothing tags, are the leaders in anti-theft technology. Between them, they are responsible for installing nearly one million anti-theft systems worldwide. They also manufacture the small anti-theft devices that come glued onto packages—the ones that set off an ear-splitting alarm if they aren't deactivated at the checkstand.

The technology used for those familiar security systems is called "electronic article surveillance" or EAS. Many of today's anti-theft EAS tags are paper-thin radio frequency tags (as opposed to radio frequency *identification* tags). They look a lot like RFID tags, but lack a computer chip and the associated privacy issues.

Anti-theft tagging is huge business. Sensormatic works with Wal-Mart as well as 93 of the world's top 100 retailers.[9] It boasts anti-theft systems installed in a staggering 445,000 locations worldwide, making it the biggest player in the EAS industry.[10] Its rival, Checkpoint Systems, runs a close second, with equip-

---

▶▶ DISABLE MY HEMORRHOID CREAM, PLEASE! ◀◀

Just about any item in the store could cause an embarrassing scene at the security gates. According to the Source Tagging Council, an industry group that cheers the practice of hiding anti-theft tags in items or their packaging, companies are now source tagging everything from condoms to corn cob pipes. Here's a sampling of some of the bizarre items we found on the organization's list:

Wallets, belts, rugs, fishing reels, ear wax product, stopwatches, golf balls, Martha Stewart flannel sheet sets, sump pumps, crock pots, cat toys, pepper spray, breast pumps, toaster ovens, book lights, personal lubricant, snoring strips, condoms, 3-D glasses, paint balls, incense burners, pregnancy tests, ear infection detectors, Star Wars toys, press-on nail kits, tents, meat soaker pads, scar removal cream, prostate formula, coffee makers, corn cob pipes, HIV test kits, deep fryers, moisture detectors, karaoke machines, and The Clapper.[13]

ment installed in 350,000 locations.[11] Their clients include well-known retailers like Target, Barnes & Noble, Circuit City, and the United States Postal Service, to name just a few.[12]

SOURCE TAGGING

In recent years, shoplifters have wised up to the presence of obvious anti-theft tags and begun simply peeling the EAS tags off before walking through the security portals. In response, security companies have deployed a new tactic: making the tags invisible by hiding them, a practice called "source tagging." Source tagging involves going directly to the product manufacturer (the "source") and getting them to embed anti-theft tags into an item's packaging or even directly into the item itself.

Source-tagged anti-theft devices are invisible by design. Checkpoint's website explains that "invisible EAS is more effective" since it is "hidden from shoplifters and employees alike."[14] So how do the companies hide the tags? By gluing them inside of sealed packages or placing them into cardboard, for example. Or they can take a more direct approach and sew the anti-theft tags into the seams of clothing or sandwich them into the soles of shoes.

FROM SOURCE TAGGING TO SPY TAGGING

Shrewdly, Checkpoint offers its customers a migration path from the low-cost, narrow pay back EAS [anti-theft] tag at a few cents to the basic ten cent RFID device.

—*ID Tech Ex*[15]

Now for the part that keeps us awake at night: Sensormatic and Checkpoint, the masters of hidden security, have begun "upgrading" their anti-theft tags to incorporate RFID. This move could spawn the creation of a broad-scale spychip infrastructure, invisible and hidden by design, emerging right under our noses. We could wake up one morning to discover that our homes are filled with spychipped items and that a ubiquitous network of RFID readers has been installed into countless commercial doorways to silently scan us everywhere we go.

This threat is very real. Checkpoint's newly-minted line of dual-use anti-theft/RFID portals (ironically named "Liberty" readers) can already read both anti-theft and RFID tags, and the company recently announced plans to purchase *one hundred million* RFID tags from vendor Matrics to meet its customer needs.[16] Where are those one hundred million tags? We'd sure like to know. Wouldn't you? They're obviously being used somehow.

While Checkpoint represented that the tags will be used for warehouse operations, the company touts itself as "the world's largest integrator of RFID technology into consumer product packaging."[17] Egads! Could Checkpoint already be hiding these things in our belongings? Might we be wearing RFID tags now and not even know it? Checkpoint and Sensormatic (which has said it is working on the same RFID schemes) aren't telling.

But as far back as 2003, Checkpoint bragged that it was "working with forward-thinking consumer product goods manufacturers and retail clients on pilots."[18] Who were the unnamed retailers that tried sending spychips home with consumers? Have their pilots now become full-scale rollouts? Were Target, Staples, or Circuit City involved? After all, they are Checkpoint clients. Perhaps it was the nation's third largest grocery chain, Albertson's, which recently announced plans to install Checkpoint's Liberty readers in its doorways.[19] Or was it CVS pharmacy, one of the nation's largest drug chains?

CVS is not only upgrading to Liberty anti-theft portals, *they're hooking the readers up to phone lines* so they can transmit data to CVS headquarters. This will "enable CVS to view alarm data by region, by store, or by date and time."[20] The reason? Management apparently wants to know exactly what products are going out the door.

Since there is no legal requirement for companies to tell consumers when products they buy contain RFID tags, it could be happening today at any one of these stores. That's why we've called for mandatory labeling legislation that would require companies to tell consumers when items contain an RFID tag and what such a tag could mean for their privacy. (More on this topic in Chapter Seventeen.)

You're not likely to carry things like Martha Stewart flannel sheets, a coffee maker, or the Clapper around with you, so you might wonder what difference it would make if these were someday laced with RFID. But what about something like your shoes?

IF YOUR SHOES COULD TALK

RFID tagging of any consumer item poses a threat to your privacy, but nowhere is that threat more obvious than in the things we wear. And you really wouldn't want to have a spychip in your shoe. The reason is simple. To understand why, ask yourself: When was the last time someone else wore your shoes? If you're like most people, the answer is probably never. For comfort, hygiene, and cultural reasons, people typically don't exchange shoes with each other. Plus, unless we're Imelda Marcos, most of us only wear a limited number of shoes. So if someone could scan a shoe and read its ID number, they might have a pretty good idea of who was standing in it.

An RFID tag in a shoe could function as a tiny spy, relaying information about your presence and movements to the readers embedded in the surfaces you walk on every day, like door mats, store floors, and the carpeting in your home. The challenge would be to find a way to slip an RFID tag into people's shoes and link the shoes' unique ID numbers with their wearer's identity. Then, a network of reader devices placed in strategic locations could make it possible to monitor the movements of vast numbers of people.

CREATING THE PURCHASE DATABASES

To understand how RFID numbers on shoes and other consumer items could be linked with you, it's necessary to understand what happens today when you shop at a national retail chain. While many people still think of cash registers simply as adding machines with cash drawers, they have become sophisticated, ubiquitously networked, high-speed computer terminals, feeding purchaser data directly into massive databases.

Unless you make a point of paying with cash (which is currently anonymous), you communicate your identity with every transaction you make. If you shop at a major chain with a frequent shopper card, credit card, or ATM card, a list of each item purchased will likely be stored in the retailer's database in a record with your name and card number at the top. Depending on the retail chain, these records can stretch back ten years or more. This means that the store can view an itemized list of purchases you've made over time.

Information aggregators like Chicago-based Information Resources, Inc. (IRI), collect the sales information from cash registers around the country and consolidate it into centralized databases. IRI has been doing it since 1987 and now claims it collects and consolidates data from over thirty-two thousand U.S. food, drug, and mass merchandise retail stores.[21] Never heard of IRI? That's exactly the problem. While some privacy-conscious consumers may wince when they hand their card to the cashier, most people give little thought to where the data goes after that. The multibillion-dollar infrastructure that captures people's personal information and traffics it to others is largely invisible.

Of course, your purchase records can then be linked up with other information about you, such as your name, address, occupation, vehicles driven, credit rating, and so on. Because of advances in geospatial satellite imaging, there could even be aerial photos of your home and neighborhood in a database somewhere.[22] So whenever you make a purchase with anything but anonymous cash, you are likely adding another brush stroke to a very intimate portrait of your life and that of your family.

Since today's cash registers, known in the industry as "point of sale" or "POS" terminals, automatically capture bar code data and record it in the purchaser's data record, we can assume that the same thing will happen with RFID tag data. The difference, however, is that the RFID tag data will include unique ID numbers. Just as today's invisible POS databases record what we buy, tomorrow's POS databases will record exactly which ones we purchase, identifying them by the unique EPC numbers encoded on their RFID tags.

Once those databases are consolidated, anyone able to obtain them would have the ability to identify people secretly, at any point where they step within range of an RFID reader. What's more, tagged shoes would make it possible to keep a record of where people have been, based on tag sightings. Not only would someone with access to such a database know, for example, that you were spotted at a retail store on a particular date at a particular time, that person might also be able to tell exactly where you were standing in the store and for how long. For example, a reader in the store floor in front of the self-help books or pregnancy test kits could record your presence, and by association, your interest in such things.

Source Tagging Shoes

So, is anyone putting RFID into shoes today? Well, we know they're putting anti-theft tags (like the ones that can be equipped with RFID) into shoes. Checkpoint slips them between the layers of the sole. Here's an excerpt from a promotional piece describing the process: "One shoe in every pair has an EAS circuit placed between the layers of the sole at the point of manufacture," explains Charlie Mills of shoe retailer Mills Fleet Farm. "Customers and employees don't know the tag even exists in the shoe."[23]

Philips Electronics, a major RFID chip manufacturer and the global corporation behind the Benetton clothes tagging project, is already thinking about tagging shoes with RFID. In a patent application filed by the company in 2003, Philips describes the need to keep an RFID device small and powerful, yet soft enough for footwear and clothing. Philips devised a fabric antenna that is "flexible and pliant, thereby lending itself to taking on and conforming to the [item's] shape."

One form of the fabric antenna is a line of "conductive threads interwoven with the fabric." To the untrained eye, the fabric antenna could look like any other stitching and it could feel just like the rest of the material of a shoe or garment. Talk about hidden!

Philips has clearly thought about reading these shoe tags while the shoes are on people's feet. The inventor observes that "the placement of [the RFID tag] in [the] shoe may be particularly advantageous where the [RFID] interrogator is located in a floor."[24]

Why would Philips want shoes to be interrogated by a reader in the floor? The next chapter may provide some clues.

▶▶ But Wouldn't You Need a Reader Every Five Feet? ◀◀

Katherine once had lunch with a reporter who couldn't see how RFID tags with a mere five-foot read range could be used to track people. "With such a short range," she sniffed, "I'd have to follow you all day, pointing a reader at you to keep you in range." She waved her hand dismissively. "Rather than go to all that trouble, I'd just watch you instead!"

But what hadn't occurred to her was that even with the relatively short read range of today's tags, snoops wouldn't need billions of reader stations to keep track of someone's travels (nor would they have to follow their target around)—provided readers were installed at strategic locations.

The reporter had driven over sixty miles that day to meet with Katherine. With one reader every five feet, a snoop would need 64,416 readers to monitor every moment of that trip. But it would hardly be necessary since the same trip could be tracked with just four readers. One reader at the freeway onramp could scan the spychip in her tires, a second at the freeway offramp could determine where she got off the freeway (and perhaps even issue her an automatic speeding ticket if she had gotten there too quickly). A third reader at the entrance to the parking lot could tell where she parked her car, and a fourth reader at the door of the restaurant could scan the tag in the reporter's shoe or read the data on her spychipped driver's license—right through her wallet.

5

There's a Target On Your Back

> If I talk to companies and ask them if they want to replace the bar code with these tags, the answer can't be anything but yes. It's like giving them an opportunity to rule the world.
>
> —Steve Halliday, vice president of technology at AIM Global[1]

Can you see him? Not if he can help it.

Bob is careful to avoid being noticed as you exit your parked car. He jots down detailed notes including your vehicle's make, model, color, and license plate number. This information will be critical to his mission. He grabs a shopping cart from the corral and follows you into the supermarket, keeping just the right distance. Not too close.

First turn, the fruit and vegetable aisle. Bob feigns an interest in peaches while you pick the perfect bananas and place them in your shopping cart. He duly notes your selection in his inconspicuous mini notebook. "Bananas, two pounds. Cost: 99 cents per pound."

You head towards the bakery. Bob follows a few paces behind, noting that you pause at a *Cinco de Mayo* display and look over the Mexican Bimbo bread.

When you add a loaf to your shopping cart, Bob imagines that you could be Hispanic—though it's not clear from your outward appearance. Bimbo brand bread is a pretty good indicator that you have roots south of the border. He keeps his eyes peeled for other telltale signs as you roll down the cereal aisle.

Since you add high-fiber cereal to your cart, Bob figures you're either (a) health conscious, (b) trying to lose weight, or (c) constipated. Bob rules out the first two options when you add Metamucil and hemorrhoid suppositories to your cart, discreetly hiding them under a copy of the Sunday newspaper you picked up before heading to the health aides aisle.

In anticipation of greeting the cheery checkout clerk, your cheeks burn though you try to compose yourself. Bob pretends not to look and updates your profile. "Clearly embarrassed. Pays cash. Total: $21.57."

Bob quickly pays for his bag of peaches and abandons his cart full of cereal and health aids near the checkout stand. He rushes to catch up with you to make last minute notes—the way you walk, your clothing, and any other details that could help him understand you, his prey.

Your eyes meet Bob's as you glance behind you before backing out of your parking space. You smile, unaware of Bob's motives. You recognize him as that young guy from the store. Bob forces a smile in return, wondering how you'd feel if you knew.

The Super Sleuth Supermarket Survey

Though it sounds unbelievable, for dozens of shoppers in middle America, this scenario really happened. The guy we called Bob could have been any one of a number of undergraduate students of Marketing 304, a retailing course offered at Eastern Kentucky University. These students were instructed to literally follow shoppers around a store, while secretly recording details about them and their shopping habits.

The assignment was called the "Super Sleuth Supermarket Survey," and its purpose was to teach budding marketing students the tricks of their chosen trade. The students were instructed to head to a store, randomly select "a stranger who is not aware that they are being watched" and develop a "profile" by secretly following that individual through his or her entire shopping trip.

While most people would expect such behavior, if discovered, to land its perpetrator in hot water, marketers don't think like most people. In fact, marketing educators saw this assignment as such an exemplary way to condition new arrivals to their ranks that this lesson plan was featured at one learning venue under the heading, "Great Ideas for Teaching Marketing."[2]

And why not? This is exactly the kind of consumer spying performed today by professionals like Paco Underhill and his company Envirosell. Underhill employs onsite researchers called "trackers" to follow shoppers around the store, listening in on them and recording their conversations.[3] Envirosell has even stooped to closely scrutinizing the behavior of customers seated at restaurants, without their knowledge or consent.[4]

Apparently, Envirosell has no shortage of clients. Fred Meyer, CVS, Trader Joe's, and Wal-Mart are among nearly fifty major retailers that have used Envirosell's surveillance services to spy on their customers.[5]

What kind of people could think that stalking strangers is not only acceptable but laudable? Who are these marketers and who gave them the right to spy on *us*?

To answer that question, we need to understand the mind of the modern marketer. Katherine is in a good position to say how the profession has changed since she received her undergraduate degree in international marketing in the mid-1980s. Back then, marketing education focused on the four Ps: product, price, placement, and promotion. The idea was fairly simple: Make a good product, price it right, put it where people can find it, and tell them how good it is. That's good, clean, honest free-market business sense. However, in the last fifteen years or so, a new P has overshadowed all the others: *People*.

The new P emphasizes the importance of knowing everything about customers in order to influence their decisions. It's also about discriminating against the bargain shoppers or the economically disadvantaged while catering to the profitable.

This isn't a bed of roses for the customers at the top, either. Their every move is under the marketers' microscope, and their valuable shopping data is analyzed for internal purposes or trafficked to the highest bidder.

Of course, the retailers, manufacturers, and marketers are savvy sorts who aren't stupid enough to say this outright in public. No, instead of riling the masses or spooking their prime customers, they discuss their plans in "market speak." One of their insider terms for spying on customers is *customer relationship management* (CRM). You can think of this term as a euphemism for consumer espionage. Bob and his classmates are learning the low-tech version of this increasingly high-stakes game.

There are many ways marketers justify CRM. They tell the public that it's for their own good—that knowing who they are and what they like to buy will help to serve them better. And, of course, since 9/11, they can pin some of their snooping on security measures that are supposed to keep us all safer. But the real objective is finding more efficient ways to separate us from our money.

University marketing students are being schooled in CRM, and thousands graduate each year to help perpetuate the new technologies that enable businesses to learn more about their customers than anyone, including your own mother, has a right to know.

DIGITAL BOB

WARNING: If Bob's supermarket spying spree upset you, brace yourself. It's mild compared to what's coming: supercharged marketers enhanced with spychip capabilities. Every RFID-enabled item will have its own digital version of Bob on standby, 24/7, waiting for an opportunity to share its unique information. The computerized versions of Bob can hide in your shoes, ride in your car, and even go to bed with you undetected.

These digital Bobs can secretly relay information using invisible radio waves whenever they're in range of a reader device. They could be bouncing inside a briefcase, cruising down the highway at sixty miles an hour, or standing on their heads and still issue their stealthy reports successfully.

▶▶ Marketing Majors Score Lowest
on Ethics Tests ◀◀

Researchers have found that marketing students score lower on measures of ethics and academic integrity than any other university majors. Unfortunately, these same marketing students will someday wield the tools of CRM to collect your personal information. "The incidence of academic dishonesty has been increasing throughout the past few decades," ethics researcher Kenneth Chapman and his colleagues explained in a recent article. "Past research has indicated that business students cheat more than their peers in other disciplines across the university. And, of particular concern to marketing educators, the current research finds that marketing majors cheat significantly more than their peers in other business disciplines."[6] Of course, these findings don't necessarily mean that all marketing students are unethical or that all marketing practices are diabolical, but it does suggest that society may want to keep a close eye on any programs planned, used, or overseen by the marketing profession.

So who or what is on the receiving end of these reports? The consortium of businesses and government entities developing the RFID infrastructure plans to send them to massive Internet databases. Once all the billions of items on the planet contain digital Bobs, theoretically, the whereabouts of everything and everyone will be known at all times and accessible to anyone with access to the databases, authorized or otherwise.

Imagine the power of being able to log onto a Google-like Internet search engine and find out all the items associated with a particular person, organization, or government entity. Then, imagine being able to find out where all those items are in real time, where they have been, and their historical relationships with other items, people, and events.

Combine that power with information from the millions of databases already in existence and you can glimpse how RFID could become the most powerful marketing—and surveillance—tool in history.

Procter & Gamble Tracks Consumers with RFID

Many companies have contributed to the RFID menace headed our way, but if we had to pinpoint a single company that bears the brunt of responsibility, it's Procter & Gamble—makers of such ubiquitous household items as Tide laundry detergent, Crest toothpaste, and Cascade dishwashing soap. As you will recall, P&G was one of the original founders of the MIT Auto-ID Center back in 1999 and should be held accountable for helping to launch the spychip revolution.

In those early years, P&G spent a lot of time and money promoting its vision to other companies, knowing that it would not be possible to create a standardized, worldwide system for tagging everything unless other major players got on board. They eventually attracted over a hundred companies to invest in the Auto-ID Center. How? By showing them, among other things, the technology's marketing potential.

Since those initial days, P&G has erased virtually every trace of its early public statements about RFID, replacing them with bland gobbledygook about "improved supply chain efficiencies" and "streamlined operational costs." But with some careful digging, we were able to snag a few of their juicier communications from circa 2000—back before the company realized it had a major consumer acceptance problem on its hands. After reading these documents, it's pretty clear why P&G might want to keep a lid on them.

The documents reveal that in 2000, P&G and MIT built a prototype "Store of the Future" and "House of the Future" in Cambridge, England, to showcase RFID's potential. They cooed about the power of RFID in a promotional piece titled "Imagine the House and Store of the Future":

The "Store of the Future" features . . .

Shelves that know what they contain . . .

Shopping carts that know what they contain. . .

Floor tiles that track products and carts around the store . . .[7]

As for the "House of the Future," it features a "TV that shows you advertising based on what you buy and what you are about to run out of, using information from your refrigerator."

If you're mortified at the thought of floor tiles tracking you or your fridge serving up content for television ads, we understand completely. Surely the "programmers of our future" (an actual quote from one of the documents[8]) weren't actually serious, were they? On the contrary, apparently they meant every word, reassuring skeptical readers that "this is science, not science fiction."[9] They even provided details about their television advertising scheme in a document titled "A Chip in the Shopping Cart":

> Let's say [an RFID tag is] positioned on the bottom of a bottle of Coca Cola. As soon as we take it out of the refrigerator, the fridge will know that the Coke supply has run out and it will inform the supermarket via Internet that they need to re-supply your home. Or, if you prefer to do things on your own, the Coke will be automatically added to the shopping list displayed on the electronic blackboard in the kitchen. At this instant, as if by magic, the publicity of Pepsi Cola will appear on the home TV screens. Because your intelligent refrigerator has communicated with your intelligent TV set.[10]

The piece wraps up with a section titled "What's in It for P&G?" Call us cynical, but we weren't surprised to see that P&G identified *improving target marketing* as one of its key objectives.

P&G's Store of the Future

It didn't take Procter & Gamble long to work out the technical details of tracking customers around its envisioned spychipped future store. In August 2001, the company applied for a patent titled SYSTEMS AND METHODS FOR TRACKING CONSUMERS IN A STORE ENVIRONMENT.[11] Subtle, huh?

In that patent application, Procter & Gamble lamented the shortcomings of existing methods of consumer espionage. Capturing people's data at the cash register and following them around the store with video cameras and human observers just wasn't cutting it:

Unfortunately, this kind of low tech approach cannot generate sufficient data. . . . [T]here is a need for empirical tools which can allow detailed analysis of what consumers experience in stores; where they go, how long they stay there, and what influences the paths they choose.

Noting the "tremendous economic incentive for both retailers of goods and the providers of such goods to understand what motivates consumers to purchase," they developed a plan to use sensors anywhere in the store, including the ceiling, floor, shelving, and store displays, to read RFID tags on shopping carts and on items shoppers pick up:

The basic parameters measured by [the] system include where an individual goes . . . and for how long. This may then be tied in with what that individual purchases based on the individual's corresponding check out or point-of-sale data gathered at check out stands. . . . [A]ctual tracking of consumers in the store environment . . . generat[es] much more substantial information . . . [that] may be used to effectively direct consumers to higher profit margin items. . . .

Using RFID to track people in the store? No, say it ain't so! After all, hasn't Sandy Hughes, P&G's "global privacy executive," repeatedly assured us that P&G has never even *considered* tracking consumers with RFID?[12]

NCR: More than Just Cash Registers

NCR is the nearly six-billion-dollar company that makes the world's cash registers. Founded in 1884 as the National Cash Register Company, NCR was one of the first global businesses, with offices today in over one hundred countries. The company now makes all sorts of other things, too, including bar code scanners, self-checkout units, and ATMs.

NCR's decisions impact a lot of people. The company proclaims on its website that "we enable companies the world over to touch millions of customers, millions of times each day." NCR's influence is felt around the globe.

▶▶ DROOLING IN THE BEAUTY AISLE ◀◀

Procter & Gamble and Giant Food Stores, part of the scandal-ridden Ahold family (the Enron-like accounting bad boys of the supermarket industry[13]) once joined forces to conduct "extensive research on consumer shopping habits." The goal was to increase sales of P&G brand health and beauty aids. According to a June 2001 article published by the Pennsylvania Food Merchant's Association, the duo paired up to "double the amount of customers that come down the aisle."[14]

Reportedly, they used a "unique methodology" to devise their new marketing approach. Marshall Haine, one of the inventors of P&G's still-pending patent to track shoppers in stores, characterized their efforts as "Pavlovian," explaining, "We tied into the consumer's basic needs—fundamental needs that inspire behavior."

Hmm. "Pavlovian"? Of course that word refers to the work of Ivan Pavlov (1849–1936), the scientist who studied ways to elicit desired behavior from his subjects. While Pavlov's subjects were dogs that he trained to drool by the mere ringing of a bell, his theories have been translated to other scenarios—now, it seems, by Procter & Gamble to get female shoppers to swarm to their cosmetic displays.

They boast that wherever a transaction takes place, whether "across the counter, by telephone, at a kiosk or ATM machine, or over the Internet, NCR is there."[15] You may have even bought something at an NCR cash register today—their client list includes such heavy hitters as Wal-Mart, Bank of America, KFC, and major supermarkets.

But in addition to selling cash registers, NCR peddles an invisible and far more insidious product: data capture. Strategically positioned throughout the world at what the industry calls the "point of sale," NCR can capture virtually limitless quantities of valuable marketing information about people and the things they buy. The company brags about how intimately it can observe customers and tells corporate clients it specializes in "giving you and your company the tools you need to gather critical data about their individual preferences, needs and requirements."[16]

It's no exaggeration. NCR is up to its eyebrows in data, telling website visitors that its Teradata division is "the market leader in data-warehousing solutions." These warehouses of data are so big they require entire buildings to house them.[17] Their size boggles the mind. The *New York Times* recently reported that Wal-Mart alone has stored 460 terabytes of data on NCR Teradata mainframes. "To put that in perspective, the Internet has less than half as much data," said the *Times*.[18]

Sitting in the catbird seat with regard to both sales and data collection, it's little wonder NCR has turned its interest to RFID—so much so that it has a full-time "RFID Evangelist" on its staff.[19] He'll tell you the company has plans. Very big plans.

CORNOGRAPHY: NCR VOYEURS WATCH WHILE YOU SHOP

In December 2003, NCR was granted a patent for an invention they call AUTOMATED MONITORING OF ACTIVITY OF SHOPPERS IN A MARKET.[20] Reminiscent of Procter & Gamble's plans, here NCR describes an unbelievably detailed strategy to watch shoppers' every move in the store aisle, recording their activities on a moment-by-moment basis and making a record of everything they do—down to the split second.

NCR's scheme starts with RFID tags on every product in the store and reader devices hidden in every shelf and grocery cart. These are in turn connected to a vast, silently watching computerized spy network that waits for a shopper to lift an item from a shelf, rather like a spider waits for a tug on its web to indicate that its next meal has arrived.

When an unsuspecting customer does lift an item from a shelf, say a can of corn, the system kicks into surveillance gear, timing precisely how many seconds the shopper holds the item before either putting it back on the shelf or placing it in her shopping basket:

> The invention monitors the items. The invention determines whether each
> item is located in one of three positions, namely, (1) in the basket, (2) on
> shelves, or (3) neither in the basket nor on the shelves.

For example, an item may take the form of Brand X canned corn. If the shopper removes a can of Brand X corn from a shelf, and holds the can in the hand, the invention will detect that a can of Brand X corn has been removed from the shelf, and also that the can is not in the basket. The inference may be raised that the can is held in the hand of the shopper at that time.... The data just obtained is recorded.

... [A] detailed record of the successive occupations of positions [basket, shelf, or neither] ... can be recorded, together with the time-of-day for each position.

Based on the recorded positions of the can of corn and the time at each position, the system makes inferences about shopper activity. For example, if the can is removed from the shelf but not immediately placed in the basket, the system could interpret that the customer was reading the label before deciding to purchase it. Likewise, if the can was put back on the shelf and a competing brand was later selected and put into the cart, the system could infer that the consumer preferred the competing brand.

This real-time shopper surveillance gives the system the ammunition it needs to sell you other brands, related products, and, preferably, higher-margin items. For example, if the system catches you buying Brand X corn, it might prompt you to try Brand Y in the future by spitting out a Brand Y corn coupon at the checkout.

NCR's system can make even more advanced inferences. For example, if a shopper places a high-end pasta in her shopping cart, it might decide she's a candidate for an expensive brand of sauce and suggest the purchase. If a shopper places a can of pumpkin in the shopping cart, "the invention may prompt the customer to purchase whipped cream, or eggs, based on the assumption that the customer intends to prepare pumpkin pie."

System inventors say the purpose of all this espionage is to "collect data on customer behavior, for the benefit of the owner of the market, and the manufacturer of the items."

Note the total absence of any consumer benefit. They must figure there's no need to even pay lip service to consumers. Why should they bother? Consumers need never even know the system is in place since all the spying is done via invisible radio waves.

ARE YOU A "BOTTOM FEEDER"? HOW ABOUT A "BARNACLE"?

"Barnacle"? "Bottom feeder"? These are just two of the derogatory terms marketers use to describe bargain shoppers.[21, 22] If you've ever been a starving college student, single parent, or laid-off worker who seeks out the cheapest deals to make ends meet, marketers might have called you one of these names behind your back.

If you're not dropping big money, stores don't want you breathing their air, pushing a grocery cart, or taking up the cashier's valuable time with your measly purchases. And if you do, they're looking for ways to make you pay.

Marty Abrams, a senior policy advisor at the influential Hunton & Williams law firm, discusses ways marketers do just that:

> At the most macro level, CRM is the process of using information technology and statistics to maximize a company's relationship with every current and potential customer. Maximization in some cases means providing white-glove service and pricing that expands the firm's share of that consumer's wallet. In other cases, it means marginal service and high prices designed to drive the unattractive consumer somewhere else. A critic of targeting — which I am not — might refer to this as digital redlining.[23]

They've got it all mapped out. The retailer's problem, however, is figuring out how to identify people so they know whom to reward and whom to treat poorly. Enter RFID.

If a store can get shoppers to carry RFID-laced loyalty cards, like those offered by Texas Instruments, the problem is solved.

TI is encouraging retailers to install doorway RFID readers for "keeping track of the customers walking in the door." Its website touts an RFID-enabled frequent shopper card that can be read right through a shopper's purse and

(PHOTO: LIZ MCINTYRE)

Texas Instruments plastic card blank with RFID tag inside.

describes how consumers "with a TI-RFID tag in their purse, pocket, or wallet can be detected by reader systems at doorways. Readout antennas can also be in counters, walls, and in floors." It also details how "the technology can tell retailers exactly who's in their store at any given moment while offering full purchase histories for each shopper. In addition, stores will know what the customer bought at their last visit and what they might need for accessories."[24]

Are spychipped loyalty cards really headed our way? Possibly. One of the world's largest manufacturers of plastic membership and payment cards, Arthur Blank & Co. Inc., has announced it is adding RFID capability to its product line.[25] This move has huge implications for consumer privacy, since the company makes 1.3 billion cards every year. You may already have a card made by Arthur Blank in your wallet, since their client list includes Holiday Inn, Barnes & Noble, American Airlines, OfficeMax, AAA, and Blue Cross/Blue Shield. They also manufacture ATM cards for major banks. Though Arthur Blank now offers the spychipped capability, their clients may be reluctant to take them up on it after what happened in Germany. We'll tell you more about *that* in Chapter Six.

MINORITY REPORT STYLE INVASIVE ADVERTISING—IS IT IN OUR FUTURE?

In the 2002 Steven Spielberg movie *Minority Report*, advertising is highly targeted and personalized to the presumed needs, wants, and desires of passersby. In an unforgettable scene, a talking display makes an apt suggestion based on the perceived state of the fleeing and agitated Tom Cruise, "Why not have a Guinness?"

Is it possible that in a spychipped future our privacy will be compromised to the point that we can be served up such insightful recommendations?

Beth Givens of the Privacy Rights Clearinghouse testified at the August 2003 California State Senate hearings on RFID that the movie was grounded in frightening factual possibilities. She reported that the movie's science and technology advisor, John Underkoffer, was directed to ensure that the technology portrayed in the film was "future reality," not science fiction.[26]

Our research into patents and patent applications leaves us agreeing with Mr. Underkoffer's technology assessment.

"IBM" MEANS "I'VE BEEN MONITORED"

We gave you a taste of IBM's patent pending invention IDENTIFICATION AND TRACKING OF PERSONS USING RFID-TAGGED ITEMS in Chapter Three and promised we'd share more of their Orwellian RFID vision for the retail store. If you recall, the crux of their scheme is to scan consumers entering retail spaces for RFID tags in order to learn their exact identities and other information about them.

Armed with this tag information, IBM proposes tracking shoppers throughout a store and using the information "in a number of different ways," including the delivery of targeted advertising, a la *Minority Report*:

> For example, if the person is carrying a baby bottle, a store advertisement system may be configured to advertise diapers while the person is passing a particular display device in the store. If the person is carrying a man's wallet, the store advertisement system may be configured to advertise razor blades and shaving cream while the person is passing through a particular display device in the store. Obviously, numerous examples are possible.[27]

IBM'S "MARGARET" PROGRAM

IBM has also developed a product called "Margaret" which uses doorway RFID readers to identify customers as they enter banks and financial institutions. Named after the developer's presumably wealthy mother-in-law, the idea is to identify valuable clients and single them out for preferential treatment. IBM describes the program as follows:

> [A]n RFID tag fitted to the customer's bank card or passbook could be used to signal their arrival at a branch. As they pass through the doors, the card would alert a customer information system. Bank staff could personally greet high-net-wealth customers, or customers could be greeted by name by tellers, who would already have their account information on-screen when they arrive at the counter.[28]

A leading industry publication, *RFID Journal*, offers up other ideas for RFID readers at building entrances, suggesting that in addition to their use in banks, "the same system could be used in upscale restaurants or retail boutiques, where a high-degree of personal service is important."[29]

Patent   Mar. 16, 2004   Sheet 13 of 13   US 6,708,176 B2

Bank of America's public advertising system "may be supported by a variety of free-standing devices ... or attached to or suspended from a nearby wall or ceiling." This figure illustrates how a user might interact with the invention. A "Focused Audio Device," "Remote Identification Sensing Device," and "Group Display Device" may all be incorporated into this display.

(SOURCE: U.S. PATENT NO. 6,708,176)

Are You Talking to Me?

Not only will you be surrounded by targeted ads at every turn in the spychipped world, they could well speak to you. *Literally.* Bank of America, one the world's largest financial institutions, tells how in its patent SYSTEM AND METHOD FOR INTERACTIVE ADVERTISING:

> [T]here is a need for a public advertising and announcement device that has the ability to identify specific individuals or groups of individuals who come into contact with the device, the ability to collect, gather and use personal information about those individuals or groups to select and present more interesting, targeted ads and announcements. . . .[30]

B of A's patent suggests scanning a transponder (that is, an RFID tag) embedded in items people are wearing and carrying like key fobs, cards, or other devices to identify them. But while you're watching the ads, they could be watching you back. The system can capture video images of consumers near the display, recording "physical characteristics, e.g., physical appearance, face, iris and retinal characteristics[.]" That data can then "be processed by a Crowd Evaluation Device or a Customer Biometrics Sensing Device. . . ."

If you think the inventors have overlooked privacy, you're wrong. They've thoughtfully designed their audio device to project only to persons who are directly in front of the display. They claim this will allow a customer to "have some degree of privacy, at least with regard to the audio program of an Interactive Poster."

Gee, thanks for the consideration, Bank of America.

6

THE RFID RETAIL ZOO

Residents of the German town of Rheinberg will find themselves guinea pigs in what Metro and its partners . . . hope will become a global standard within the next five to ten years.

—IBM News Release, April 28, 2003[1]

TOURING THE STORE OF THE FUTURE

If there's one place the RFID industry should be on its best behavior, it's the METRO Extra Future Store in Rheinberg, Germany—especially if CASPIAN and German privacy organization FoeBuD are stopping by to tour the store.

The Future Store is the item-level RFID tagging showplace of the world. There, companies like IBM, Procter & Gamble, Gillette, NCR, and Kraft are running RFID experiments on real shoppers in a real store. Spychipped bottles of Pantene Shampoo, packets of Gillette razor blades, and tubs of Philadelphia brand cream cheese sit on RFID-enabled shelves, waiting for consumers to take them home.

Future Store executives agreed to show us around their living laboratory. While everything was cordial, there was no mistaking the tension in the air. METRO really wanted our seal of approval, but things didn't quite go their way.

The execs assured us that the RFID tags on these products posed no risk to consumers since they could be disabled at a deactivation kiosk after purchase. While we were told the kiosk would erase information on the RFID tags, we discovered that even after a product had been through the so-called deactivation process, its serial number could still be read from up to five feet away.

That was bad enough, but the real find came when Katherine went to get a METRO "Payback" frequent shopper card for her collection. She has been fighting shopper card schemes for years, and they were the subject of her doctoral thesis at Harvard. But getting the card wasn't easy. The store employees weren't authorized to simply hand out the cards and had to wait for approval from upper management. While this struck us as strange at the time, we didn't find out the real reason for their reluctance until the following afternoon during Katherine's talk on privacy in Bielefeld, Germany.

When she finished the slide portion of her lecture on the dangers of RFID, FoeBuD's co-director hooked up a 13.56 MHz RFID reader to the laptop she was using and projected it onto a screen behind her. One by one, she and a colleague held spychipped products they'd purchased from the Future Store up to the reader device so the audience could see the data encoded on the chips.

Then came the extraordinary moment when someone picked up the Metro Payback loyalty card and held it to the reader. Nothing was supposed to happen. But then a string of numbers appeared on the screen! Our jaws hit the floor. There was a spychip in there. We'd been tagged!

Spychipped Loyalty Cards

In retrospect, we shouldn't have been shocked to find a hidden RFID tag in Metro's loyalty card, considering who was behind the Future Store.

IBM, with its extensive plans to track people with RFID, was a key player in the store, "providing the overall systems integration, [and] the RFID middleware."[2] We should have known something was up when IBM referred to the people of Rheinberg, Germany, as "guinea pigs" for their Future Store experiment.[3] Then, there was the involvement of Procter & Gamble and Gillette, the patron saints of the spychip, with their shared history of watching unsuspecting shoppers with hidden RFID tags. No good could come of that alliance.

But the real tip-off should have been the starring role played by the retail snoops at NCR. Just a few months earlier, we had obtained a copy of an NCR promotional piece called "50 Ideas for Revolutionizing the Store through RFID," a forty-six-page, full-color extravaganza that laid out a store full of invasive plans to identify shoppers, track their movements, manipulate prices, and more.[4] Many of these plans hinged on embedding spychips into loyalty cards.

Some of NCR's fifty ideas were technical, such as No. 7: "Merchandise leveling across stores" (yawn), while others could make shopping more efficient, like No. 23: "RFID enables checkout without removing items from the shopping basket/cart" (the "killer ap" they hope will convince you to embrace RFID). But mixed in with these practical and whiz-bang applications were others so disturbing they gave us the creeps.

Sometimes you have to laugh to keep from crying. So rather than report NCR's plans with a straight face, we thought we would give these ridiculous ideas the send-off they deserve. You may remember the 1970s Paul Simon hit "50 Ways to Leave Your Lover." We've updated a few of the lyrics and rededicated the song to the resourceful NCR spychippers working overtime on ways to watch us all.

> *They said it's now become*
> *their habit to intrude,*
> *nonetheless they hope their snooping*
> *won't be seen or misconstrued. . . .*

♪ "JUST USE IT TO SPY, SLY" ♪

NCR foundational idea: *"[I]ndividual shopper movements through the store could be precisely tracked in real time."*

NCR devotes a whole section of their fifty ideas document to "people tracking," a way to turn consumers into exhibits in a bizarre retail zoo. Thanks to

RFID, their surveillance tools can be blended invisibly into the environment so consumers will behave naturally, never knowing they're being watched or manipulated by their hidden keepers.

NCR knows the easiest way to track customers is to get them to carry an RFID-enabled device that can be read silently at the door and then used to pinpoint their locations as they shop. A loyalty card would be ideal—and, as we've seen, RFID proponents have already tried using them to experiment on unwitting shoppers.

Once the store has the tracking device safely tucked into your wallet, NCR suggests they work their way up to full-scale observation by gradually adding reader devices to the store environment. "As readers show up in more store locations for one function (e.g., employee identification), they can potentially be used for other functions (e.g., shopper identification and merchandise tracking)."

NCR's document even has a chart listing all the prime places for "people tracking." It includes the parking lot, entrance, exit, snack bar, aisles, point-of-sale, customer service area, pharmacy, photo desk, deli, bakery, video area, layaway, fitting room, auto center, and—well, just about everywhere.

♪ "Mess with the Price, Bryce" ♪

NCR Idea No. 34: "*With RFIDs on loyalty cards to identify the customer and a customer shopping history database, items could be priced differently depending on characteristics of the person who was buying them.*"

Imagine approaching a shelf and seeing the price tag change before your very eyes, flashing you a personalized price tailored to your shopping history and profitability to the store. It's called "customer specific pricing," and spychips could make it a reality.

You know that awful feeling you get when you sit next to a guy who paid $100 for the same flight that cost you $600 to board? In NCR's retail zoo, you could have the same experience with food, clothing, and even children's toys, every time you shop.

If you're thinking stores plan to charitably offer lower prices to the poor, we've got bad news for you: They intend to do just the opposite. Stores want to reward their best (that is, *most profitable*) customers with discounts and special perks to encourage their loyalty, while forcing low-profit shoppers to cough up more dough by charging them higher prices.

Online bookseller Amazon.com tried charging consumers different prices for the same items—that is, until they were caught in the act. One savvy customer noticed that when he deleted the cookies from his computer (those small bits of software code that can ID website visitors), the price of the DVD he wanted to buy dropped from $26.24 to $22.74. Amazon had apparently used the cookie to identify him as someone willing to pay higher prices than others, based on his past behavior. Public outrage over the personalized pricing scheme forced Amazon to discontinue the test and offer refunds to the customers they'd fleeced.[5]

Though Amazon got caught, other online merchants today continue this practice in secrecy. Traditional brick and mortar retailers would like to do the same thing, but there has been no physical equivalent to the cookies—until now. NCR knows that putting spychips into loyalty cards would supply just the tool retailers need to make price-changing shelves a reality. By embedding an RFID tag in your loyalty card, the store can ID you from the moment you walk through the door and set prices for your entire shopping trip.

♪ "Get 'Em to Wave, Dave" ♪

NCR Idea No. 26: *"Credit/debit card and customer loyalty information could be placed on a single card, allowing shoppers to take advantage of promotions and provide payment with a single card reading."*

Don't leave home without it—your RFID chip, that is. American Express, Visa, and MasterCard have begun embedding RFID tags into their credit cards, in a move that will enable even more detailed information to be gleaned from consumers wielding them. With just a wave of their wrists, shoppers can transmit payment information just like they currently do with the now familiar Mobil Speedpass.

In NCR's dream world, customers would also be transmitting information about their purchase history and value to the store that is programmed into the card's microchip.

(PHOTO: PETER EHRENTRAUT/FOEBUD E.V., GERMANY)

Katherine Albrecht examines shelves containing tagged bottles of Procter & Gamble's Pantene shampoo. An RFID tag is affixed to the bottom of each bottle.

Front of Metro Payback loyalty card. There is no mention of the RFID tag embedded inside, nor did METRO provide notice to customers about the RFID tag on the card application, store signage, or elsewhere.

(PHOTO: KATHERINE ALBRECHT)

Back of Metro Payback card. There is no mention of the embedded RFID tag here, either.

(PHOTO: KATHERINE ALBRECHT)

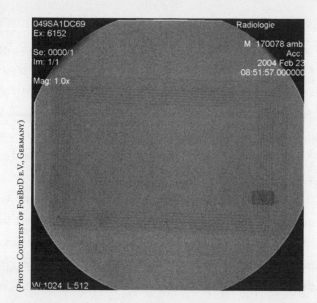

(PHOTO: COURTESY OF FOEBUD E.V., GERMANY)

049SA1DC69
Ex: 6152

Se: 0000/1
Im: 1/1

Mag: 1.0x

Radiologie

M 170078 amb.
Acc:
2004 Feb 23
08:51:57.000000

W 1024 L 512

X-ray of the Payback loyalty card confirms the RFID chip embedded within it.

NCR must have done cartwheels when the Real ID Act passed in the spring of 2005, federalizing control over driver's licenses. The act gave the Department of Homeland Security the power to set technology standards for licenses—including the potential to require them to carry RFID chips. Requiring spychips in licenses would mean consumers could not leave home without a tracking device, at least not if they're driving. (This bill has been widely denounced by civil libertarians as creating a de facto national ID card.)

Credit card companies probably celebrated right along with NCR since consolidating identification and tracking documents has been on their wish list for a long time. As far back as 1998, Alan Glass, a senior vice president at MasterCard, looked forward to the day when "[a] senior citizen could have securely protected medical information, supermarket loyalty programs, social club membership and access, discount programs, a municipal transportation pass, and a library card all stored on a single chip."[6]

Transmitting data with a wave? More like a tidal wave if you ask us.

♪ "Watch 'em Walk, Hawk" ♪

NCR Idea No. 9: *"Store cameras could be programmed to automatically pan and tilt to follow the customer with [RFID-tagged] merchandise until it is paid for."*

Yes, you read that right. Unless you'd like your every move closely watched by "pan and tilt cameras," you'll want to stay away from thirty-dollar bottles of perfume. You'll want to avoid buying more than one pack of razor blades, too, because as soon as these items carry spychips, NCR wants them used as homing beacons to guide surveillance cameras to watch your every step.

Already those cameras have become more sophisticated than you might think. No longer limited to merely watching the fifteen or so feet in front of them, the cameras can now follow individuals by handing off their images from camera to camera to obtain a seamless video of their entire shopping trip.[7]

NCR describes how RFID tags would enable store employees to closely "follow the movement of the merchandise" (of course, by extension, this means *you*) with a network of cameras that would tail you from the moment you pick the items up through the remainder of your shopping trip. If you recall, Gillette has already used a similar type of photo surveillance triggered by hidden spychips at a Tesco store in Cambridge, England. This plan works with or without a tagged loyalty card.

The problem with watching every person who picks up a given product is that it relies on a "guilty until proven innocent" anti-theft strategy. Under NCR's plan, everyone is presumed to be a shoplifter until they prove otherwise by paying for the goods. We believe that people buying high-ticket items should receive the store's thanks, not its Sherlock Holmes routine. Save the scrutiny for the bad guys—and leave the rest of us alone.

♪ "Dump the Receipt, Pete" ♪

NCR Idea No. 25: *"The receipt would be 'on the tag' in the sense that the information usually appearing on a paper receipt (and probably more information) would be associated with the RFID code for the item."* And No. 29: *"Returns and exchanges with digital receipts. If an item had been purchased using RFID, returns and exchanges could be done without paper receipts, because the information usually needed from the receipt would be associated with the RFID [tag]. . . ."*

Under NCR's "digital receipt" plan, shoppers will have to leave *live* RFID tags attached to their purchases. Not only does this mean that stores won't kill the tags at the checkstand, but if privacy concerns prompt you to remove or disable a tag once you get home, you could be destroying your proof of purchase and possibly voiding your warranty, too.

As for returns, stores can't wait to encode purchaser information directly onto products so they can limit the number of items a customer can return or exchange. Several national retailers including KB Toys, Sports Authority, Staples, and trendy clothing store Express have already begun monitoring customers' return and exchange activity through sophisticated databases.[8, 9]

If a shopper surpasses her so-called "return allowance" in a given time period, these stores may prohibit any future returns, even if the merchandise is defective. Industry numbers suggest that Express, for example, may now hand 1 to 2 percent of its shoppers a slip of paper reading "RETURN DECLINED" instead of giving them cash or credit slips.[10] While shoppers can get around these restrictions by asking friends to make an exchange for them, in the spychipped future, the item's RFID tag could be encoded with a "do not return me message," regardless of who brings it back. And that's just how the stores want it. If you exchanged the lime green bath towels your Aunt Clotilda gave you for your birthday (and everyone else did, too), you can forget about returning the fur pillow shams she buys you at Christmas. By then, your aunt would have exceeded her purchasing "error allowance."

Return Declined.

♪ "PUT THE SQUEEZE ON ST. NICK, RICK" ♪

NCR Idea No. 32: *"Dynamic pricing. RFID can be used in conjunction with electronic shelf labels to automate pricing based on the number of items on the shelf. . . . For example . . . when certain popular items were in short supply (e.g., at Christmas time), the price can be automatically raised."*

Bah, humbug! Who thinks these things up? The next time someone in the industry tries to tell you retailers only want RFID "to improve your shopping experience," you can ask when they hired the Grinch and Ebenezer Scrooge as strategy consultants. With this scheme in place, of course, you'll always wonder if that Tickle-Me-Elmo's price tag reads $589 because it's Christmastime or because they jacked up the price when they saw you coming.

♪ "Remind Her to Pay, Jay" ♪

NCR Idea No. 37: *"If a shopper had outstanding business to settle with the store (e.g., overdue or soon due video return or upcoming payment on a layaway item), the shopper could receive an automatic reminder upon entering the store. This could be done via cell phone . . . if that device had an RFID tag that identified the shopper or if the shopper carried an RFID-tagged loyalty card and the store had the customer's mobile contact information available."*

If we understand correctly, this means whenever a shopper walks through the door, the store could start calling her RFID-laced cell phone to nag her. We can just hear the annoying computer system now: "Hello, Mrs. Mac-In-Tire. Your video fine now totals six dollars and zero cents. And we have a special on your favorite dandruff shampoo in aisle three."

But the cell phone could be used for a lot more than that. Cell phone manufacturers already envision using the phone as a sort of souped-up Mobil Speedpass on steroids, where the phone contains an embedded RFID tag that is paired with the phone's location tracking and communications potential. The cell phone could ID you to merchants, flash you with marketing and advertising messages, and let you make a purchase virtually anywhere on the planet—even miles away from a cash register. (If this happens, expect cash registers to disappear along with cash.)

Such phones are already being used in Japan and parts of Europe. One recent trial in Japan even used the phone to pay cab fares.[11] (Will we soon be saying *sayonara* to anonymous cab rides, too?) At some point, retailers even hope to use mobile phones to flash prices at us when we shop and do away with the shelf price tag altogether. Companies envision a day when you wouldn't be allowed to enter a store at all without a cell phone or similarly equipped personal digital assistant like a Palm Pilot that would identify you, track you, and set prices for you as you walk around.[12]

♪ "ENFORCE A FOOD BAN, STAN" ♪

NCR Idea No. 41: *"Warnings about contents to which a shopper or family member is allergic or wants to avoid. If food or clothing were tagged with RFID that could provide information about the ingredients and materials composing the item, shoppers could be warned about items to which they or a family member was allergic when those items were placed in their shopping carts/baskets equipped with RFID readers. This would be done by having software that compared the contents of the selected items with profiles that the shoppers set up. . . . A smart system . . . could suggest alternatives that did not contain the problematic component and tell the shopper where to find them."*

This sounds innocent enough, and may even be helpful to some people. The trouble is that in order to make it work, you'll have to let the industry put RFID tags on every item in the store as well as a reader in your shopping cart. You'll also have to identify yourself on each shopping trip and let the computer review and approve (and record) every item you buy. It's a marketer's dream come true—and precisely why they'd love to sign up people for this service. After all, if you *ask* them to closely scrutinize your purchasing behavior, you can't blame them later, right?

But that's not all. Imagine what would happen if health insurers, public health officials, and even employers could also peer over your shoulder at your food choices and set their own restrictions on what food and products you could or couldn't buy. Already, police departments have fired officers for smoking cigarettes in their off duty hours, claiming that smoking raises health insurance costs for others.[13] Starting in 2007, employees in King County, Washington (the county that encompasses the Seattle area), will be charged an extra one thousand dollars in annual healthcare fees if they don't participate in a snoopy "healthy incentives" program that monitors their lifestyle choices.[14]

Why wouldn't these same tactics someday be deployed at the supermarket? NCR's Bad Idea No. 41, the grocery cart that watches your spychipped food choices, would make it possible for employers and health insurance companies to impose similar conditions on people's grocery store purchases. Why stop at tobacco? Cops could lose their jobs for buying red meat or beer. King County could go one step further and require employees to join the "healthy grocery cart" calorie restriction program or face hefty health insurance surcharges.

Giving computers the power to prevent shoppers from buying certain products sounds like a Big Brother increment just waiting to happen. Frankly, we'd rather read the labels ourselves than put a sophisticated food control network into the hands of marketers and giant corporations.

BACK TO THE FUTURE STORE

Future Store offers an unequaled glimpse of what the marketplace will look like in the years ahead, and its value as an educational tool cannot be overstated.
                —Tracy Mullin, president and CEO, National Retail Federation[15]

It's true that a lot of education takes place when retailers run trials of RFID. But while the industry may have thought their lessons would all center on read rates and frequency ranges, they were unprepared for the "F" they got from consumers who learned they'd been spied on.

Any retailer foolish enough to buy into NCR's spychipped retail zoo concept could learn a lot by looking at how consumers responded to being spied on at the Metro Future Store. By implementing *just one* of NCR's invasive ideas—the spychipped loyalty card—the company brought down the wrath of practically the entire German nation. After seeing evidence of the hidden RFID tag, the German people wanted to see the Future Store shut down.

Metro was forced to recall all ten thousand cards it had issued to unsuspecting shoppers and was pummeled with a protest outside the store. The chain received crushing negative media coverage in almost every news outlet in Germany. Even the government got involved, putting the store and its partners under investigation for potentially violating privacy laws.[16] It was clear that the

German people were not going to take hidden "Schnuffelchippen" (their word for spychips) lying down.

Our German colleagues from FoeBuD appeared on dozens of television news programs, gave testimony to government officials, and were interviewed in all the major papers. Even today, Google lists over forty thousand stories on the scandal. We agree completely with the National Retail Federation that the Future Store's "value as an educational tool cannot be overstated."

The consumer backlash against the hidden RFID chips in the Payback loyalty card sent a clear message to the rest of the world: We consumers have the power to stop RFID cold in its tracks, and we don't have to put up with retailers' nightmare vision for our future.

Here's betting NCR's "50 Ways to Use RFID" will have the spychippers quickly singing "50 Ways to Lose Your Customer" instead.

Even though the town of Rheinberg was hit with a snowstorm on the morning of the protest, angry consumers voiced their concerns. The orange sign cleverly contains the word "perfide," which means the quality or state of being disloyal—treachery.

German consumers gather outside of METRO's "Extra Future Store" in Rheinberg, Germany, to demand an end to RFID experiments taking place there. Because of the consumer backlash against the hidden RFID chips in the Payback loyalty card, the store was forced to recall ten thousand cards.

7

BRINGING IT HOME

[I]f you let them, companies like Gillette will monitor personal use of their products [in your home]. Throw one of their razors in the trash, and another one would be on its way.

—Charlie Schmidt, 2001[1]

TRAPPED BY A HOUSE PLANT

" I'm doing very well today, thank you, Jimmy," Edna rasped, her pink slippers making their characteristic scuffling sound as she moved from her bed to the window blinds.

"There now. Let's shed some light on this, why don't we," Edna said with unmistakable sarcasm as she twisted the rod with her arthritic fingers to open the blinds, hoping to beat Jimmy to the recommendation. She hated the cheery reminder Jimmy uttered if she failed to make her room "happy with sunshine" by 8:00 A.M.

"Don't you love the sunshine!" Jimmy cheered as the morning sun beamed through the vinyl slats. "Time for vitals."

As Edna scuffed into the bathroom, the medicine cabinet powered up. "Good morning, Edna. You look well this morning," the mirrored contraption monotoned. There were only two things Edna liked about the cabinet: It didn't fake a personality like Jimmy, and it was pretty easily tricked.

Edna placed her arm in the cabinet's blood pressure cuff as required every morning and faked a smile for the mirror, revealing ninety years of hard-won wrinkles. Then, as she did almost every morning for the last three years, she stuck out her tongue with a giant "ahhh," and then donated the contents of her bladder and bowels to the beckoning smart potty. She learned from past experience that exhibited unhappiness and failure to provide vital signs for transmission were too complicating.

"Time for meds," the cabinet bleeped.

This was the part of the routine Edna enjoyed the most. She opened the pill bottles, removed the prescribed number of pills, and faked taking them for the mirror. Following with a dramatic swig of water capped her bluff.

She knew better than to take the myriad of chemicals prescribed by her physician and mandated by her well-meaning sons. The medicine made her feel achy, numb, and powerless. By contrast, committing the tablets to the cleavage of her pink pajamas for sneaky disposal gave her a rush that imbued her translucent skin with a rosy glow.

After her performance, she scuffed with a bit more attitude towards the coffeemaker. Sipping coffee while poring over the morning paper was a ritual that made even Jimmy's interjections bearable.

"Your blood pressure is a little elevated this morning. We will not have coffee," Jimmy said matter of factly as Edna's RFID wristband got within read range. "It could raise your blood pressure."

Edna glared at the idle Mr. Coffee on the shelf next to Jimmy. No amount of clicking the on and off button made any difference. Her morning passion had been nixed by the unseen, all-knowing wisdom that ran the three rooms of her world and locked her out of danger for her own good.

Edna wanted to strangle Jimmy or, at the very least, rip off a few of his leaves. But she knew it would be of no avail. There would be a replacement plant and some unpleasantness like the other times. At least it was a "he"—well, the voice was a "he" voice. Much better than the strident "Judy" she had dismantled the last time.

Unaware of its peril, Jimmy launched into his morning questions. "What day of the week is it, Edna?"

"I hate you, Jimmy."

"Oh, Edna. You're so funny," Jimmy laughed. "That's why we're such good friends."

"You're a mechanical son of a bitch, if I've ever seen one."

"Oh, Edna, You're so funny," Jimmy laughed again. "That's why we're such good friends."

"Jimmy, if I knew where you came from," Edna fumed, "I'd send you back. Where *do* people buy things like you?"

"Oh, Edna. You're so funny . . ."

Goodbye to the Human Touch, Hello Big Brother

Where would one get such an obnoxious companion plant for the elderly? A company called Accenture, an international technology consulting giant, has already developed a prototype of such a houseplant:

The Caring Plant listens as an elderly or ill person talks to himself or someone else, picking up on even subtle groans or sighs. . . . For professional care providers as well as family members, the Caring Plant can provide on-site, around-the-clock monitoring, serving as the eyes and ears of caregivers.[2]

To fully understand why anyone would dream up this plant and other invasive devices on the drawing board like Internet-enabled medicine cabinets and smart potties, you have to get into the mindset of companies developing them. Global players like Accenture, Intel, and Philips Electronics are already creating homes where "intelligent objects will be able to act based on the input gathered by technologies such as web cameras, sensors and radio frequency identification

(RFID) tags, which can collect, store and transmit basic information about objects and the environment."[3]

They expect gadgets like the Caring Plant to be a gold mine because of the forecasted doubling of the world's citizens aged sixty and over by 2020.[4] This graying of the population means more people will need nursing-home care, assisted living, and some companionship in their final days. Rather than encourage family members to provide this care and support themselves (where's the profit in that?), these companies are promoting pricey technology to do the caring instead. To get the elderly on board, they'll use the proven carrot and stick. Studies show that ninety percent of elderly people would rather age in their own homes than move into "assisted living communities"[5] (and who could blame them?), so the veiled threat is they must either accept the invasive technology or get stuck in a nursing home.

Accenture is working overtime to develop artificial intelligence systems that can keep an eye on people and brags that high-tech in-home surveillance devices will not only be equal to the best human caregiving but will even be superior at getting the patient to behave in desired ways. They explain, "[W]hile caregivers and healthcare providers may try to influence good behavior, their reminders tend to be sporadic and easy to dismiss. Technology, however, can provide continuous motivation that is both subtle and powerful."[6]

But why limit such powerful technology and its electronic "whip" to grandma and grandpa when it has so much commercial potential? Accenture believes the whole family should get in on the fun: "Beyond the elderly, technology . . . could open a new set of services that bring retailers into people's homes."[7] Accenture, for instance, has a patent pending on a wardrobe that tracks all the clothing in your closet. The benefit for you is that it will suggest coordinating outfits; the benefit for them is that "the wardrobe closet may also be connected to websites via the Internet . . . to determine the user's clothing needs and find clothing offered for sale at one or more web sites." The user is then encouraged to purchase the clothing.[8] Such technological applications open up "more channels for marketing products—for example special offers on products that retailers know to be suitable. . . ."[9] Of course, they can only "know" these products to be "suitable" if their technology is continually spying on your belongings and reporting back to marketers.

The idea of marketers observing what we do in the privacy of our homes is offensive, but the thought of the government watching us is downright scary. That hasn't stopped Accenture from suggesting that government agents could use their RFID tools to monitor people in their homes—as a benevolent gesture: "[G]overnment agencies will also be looking for new ways to ensure the well-being of the people they serve," they note in a chilling comment. "Activity-monitoring tools could give [government] caseworkers a powerful complement to home visits," they suggest.[10]

We suspected that this was how Accenture ultimately aimed to cash in with their home monitoring system—to force feed us advertising and allow Big Brother to monitor our daily activities.

WATCHING YOUR EVERY MOVE IN THE HOME

Other companies are developing eerie in-home people-watching systems, too. Plans are afoot to rig homes with cameras, microphones, breathing monitors, proximity sensors, and RFID devices à la *1984* that can sense not only where occupants are in homes, but even what they're thinking and feeling.

One approach to watching people involves observing the motion of RFID-tagged items. Intel researchers call this the "invisible man" approach, after the 1933 movie classic in which an unseen person's behavior is inferred by the movement of objects with which he interacts. For example, when a cigarette hangs in midair and smoke wisps out from it, the audience understands the invisible man is smoking.

Intel's "invisible man" system rigs the home's occupant with "an unobtrusive RFID reader" that infers his activities from what he does with various tagged items: "For example, the system might detect the motion of the medicine cabinet door, followed by the motion of the bottle of vitamin B tablets, followed by the motion of the water glass" to know that the occupant has taken a pill.[11]

The Intel team looks forward to the day when spychips appear on everything, so they will no longer have to apply them by hand. "As RFID for supply chain tracking becomes more common, it may not even be necessary to place the tags on some of the household items—they will already be there," they enthuse.[12]

Other "smart homes" will rely on video cameras and microphones to do the watching. Philips Electronics takes home espionage to new extremes in their patented AUTOMATIC SYSTEM FOR MONITORING INDEPENDENT PERSON REQUIRING OCCASIONAL ASSISTANCE.[13] Who doesn't need a little help every now and then?

Their system would capture video images round the clock and even keep track of the color of your clothes (presumably, wearing Goth black might tip them off to a somber mental state), the pitch of your voice, the way you gesture, and the number of times you "enter and leave a single scene," with the goal of determining your "mental state/health status." Equipping the home to hear its occupants is a priority for the Philips team. Audio monitoring could listen to conversations and home sounds to keep tabs on the home's occupants and visitors. No sound would elude its sensitive microphones, not even the faintest click, sigh, or intake of breath.

Should you slip up and let on that you are unhappy living in such a home, the system would kick into overdrive. If it detected "the audio signal generated by a crying person," for example, the video monitors would capture your tears on tape so your captors (a.k.a. caring family members) could view them later.[14]

But Philips wants to do more than watch, of course. They want to sell you stuff. In fact, the company secured a patent in July 2004 with which they hope to fully exploit hidden RFID information on items in our homes for direct marketing purposes. According to the patent, the company looks forward to "an environment in which inexpensive machine-readable label devices [a.k.a. RFID tags] . . . appear in a great variety of contexts, as do bar-codes presently." Philips is especially looking forward to spychips appearing in Cap'n Crunch cereal—a brand they actually mention in their patent documentation.

To fully appreciate Philips's patented marketing concept and how it ties in with Cap'n Crunch, you must imagine you have purchased a table saw equipped with an RFID reader in your basement workshop. It's the latest model, hooked up to the Internet. You paid a fortune for it since it lets you scan different building materials and get instant information, like which saw blade to use on Formica, or perhaps plans for constructing a cabinet. You kick the tires, test the saw on a few boards, then, feeling hungry, you head to the kitchen for a snack. You return with an RFID-tagged box of Cap'n Crunch cereal, toss a crispy handful into your mouth, and set the box down.

On the table saw.

Oops.

Soon, courtesy of Philips Electronics patent No. 6,758,397, you start getting the weirdest junk mail addressed to "the snack lover at 533 Maple Lane," including cents-off coupons for Froot Loops tucked onto a trial subscription to *Modern Woodworker* magazine. How could junk mailers possibly have known what you were doing in your basement? Philips explains:

> The smart appliances are all network-enabled, meaning they each have a microprocessor and at least an input or output device to communicate. . . . For example, the table saw . . . may be equipped with a fixed reader . . . [that] receives data from [an RFID tag] and transmits this data, along with an identification of a user. . . .
>
> [Suppose] the user was eating cereal as a snack while working in his/her tool shop . . . a cereal manufacturer would probably have information about cereal (or other products that could be cross-sold) that is particularly relevant to users who like to eat cereal as a snack . . . suppliers of content to readers can exploit the "hidden" information in requests for information for directed marketing.[15]

This wireless marketing scheme is certainly not limited to table saws and Cap'n Crunch, so swearing off the combination of woodworking and cereal would be pointless. Philips plans to capture just about any tag information it can siphon through Internet-connected objects that "phone home" when you least expect it.

Companies like Philips also have visions of washing machines that monitor spychipped clothes, microwave ovens that read tags on frozen entrees to automatically set cook times, and even refrigerators that keep tabs on what you're eating.[16, 17] In fact, monitoring your family's food consumption is a recurring theme for the spychippers. UC Berkeley adjunct marketing professor Peter Sealey predicts that some day this ability will allow companies to "precisely target individuals" with personalized ads: "[W]e could measure whether we delivered the commercial to you, and, as I am monitoring your pantry, whether you bought the product, too."[18]

It has been reported that Gillette's vice president of global business management, Dick Cantwell, looks forward to "using readers to track consumer use of its products at home. Gillette sees the technology engaged in direct consumer marketing, which would rely on personalized information obtained from readers installed where products are actually used—in your refrigerator, say."[19]

Would people actually be crazy enough to install RFID readers in their own homes so marketers could watch them? Industry proponents are betting they can get people to submit to this type of round-the-clock food surveillance through the promise of "safer, faster, more efficient food recalls."[20] Presumably, the argument goes, if a batch of tainted peas gets recalled, health officials could quickly identify the precise location of every can in the nation, tracing them right into people's kitchens and onto the shelf.

This promise may sound good to some, but it's unlikely to actually work out that way. The food retail industry once promised the same thing with supermarket frequent shopper cards, but even though food retailers have now amassed detailed records on the food purchases of over half the households in America, we can't name a single instance where those records were used for a recall. Instead, the benefit has all gone to the retailers who collect, sift, and sell our food purchase data for their own ends.[21]

The home-watching vision doesn't stop with monitoring the pantry either. Companies actually want to see RFID readers built directly into homes to monitor everything all the time. Gillette's Cantwell says he looks forward to the day when RFID readers are "going to be a ubiquitous part of construction, whether you're building stores or homes. . . ."[22]

Already, Armstrong, maker of floors, ceilings, and cabinetry, sells ceiling tiles that come pre-embedded with hidden antennas.[23] While today those antennas might be used for wireless Internet access, tomorrow they could be part of a pervasive RFID home-monitoring scheme. Another company, Wisetrack, also sells a line of "environmentally unobtrusive" RFID readers designed for interior use. Perhaps the strangest is the RFID antenna hidden inside a picture frame that the company boasts "may be mounted unobtrusively on interior walls."[24] Who would ever guess that the picture they were admiring could be looking right back at them—and scanning the RFID tags in their shirt and shorts?

Inventory My Entire Home, Please

Remember how we opened the book with the scary idea that someday every item in your closet could be numbered and tracked remotely? We weren't joking. It's not only possible, but it's all gussied up and documented and ready to be patented. This doozy of a patent application describes how RFID readers called "sniffers" would be placed in your home's doorways, floors, and closets, and even in your car to inventory your spychipped possessions and send the results to marketers. Here's how it works:

> As the customer enters the door to his residence, a sniffer placed on the floor near the doorway detects the new [RFID-tagged] purchase. In a preferred embodiment, this wireless sniffer automatically and continuously emits an interrogation signal that searches for an RFID label which it has never seen before. The user's house may contain many sniffers, which all communicate wirelessly with a personal computer. A [mobile] sniffer could even be installed in the user's car. . . . [T]his mobile sniffer would be able to report new purchases as the car enters the driveway or garage.

As with nearly everything else related to RFID, the motive is to spy on you for marketing purposes (though we bet the government could think of some stellar uses for this scheme):

All the items in the customer's house are attached to RFID labels, which are automatically detected by a sniffer or sniffers placed inside the house. A personal computer inside the house keeps track of the inventory of items in the house, and periodically reports the inventory automatically to the retailer via a modem using a conventional telephone line. The retailer and/or his suppliers use this information to analyze their sales and marketing strategies.[25]

Not only can the sniffers watch what you take into the house, but they'll know what you remove as well, since the front door sniffer can be configured to sense RFIDs as they pass out of the house, too, according to the patent.

We have this crazy image of people someday hauling their purchases in through the windows of their homes to sneak them past the marketers. Okay, you're right, it's a ridiculous idea—since they'll probably have the windows wired, too.

Positive ID

Of course, there's one more element that is necessary for planned pervasive in-home monitoring. We must be willing to identify ourselves to the system. After all, these smart appliances cannot respond individually to users and adapt to their needs—or collect information about them for marketing purposes—if they can't seamlessly and consistently determine who those users are. Concerns over identification permeate researchers' work, amounting to a near obsession.

If you can't tell who's taking the prescription medication, who's drinking too much coffee, or who might need some psychiatric help because he's sighing a bit too much, what good is all the spying? Now that we have the ability to do it, the pressure to require some form of permanent, foolproof ID is bound to follow.

Researchers have identified a need for "natural" ways to track home occupants. The easy methods—external badges pinned on a shirt or swinging from a tether around the neck—are not ideal. Not only are they a constant reminder of the intrusion of the technology, but the tracked individual must make a conscious decision to wear them.

Even if a home's occupants agree to participate in a home surveillance system, there are times when badges would be uncomfortable or impractical. Georgia Tech researchers found "users will not wear the badge while sleeping (in order, for example, that the house can identify them when they arise to use the facilities in the middle of the night), or while doing work in the yard." The trouble is that users have to remember to put the badge back on when they get out of bed or reenter the house—unlikely at 2 A.M. or after mowing the lawn and in desperate need of a cool lemonade. "In addition, adding new users, such as frequent visitors, requires another physical badge or tag," they bemoan.[26]

It's not inconceivable that if we succumb to the lure of the spychipped home and 'round the clock identification becomes necessary for doors to open, security systems to deactivate, and appliances to operate, we might be tempted to consider the most extreme form of identification. When external badges become annoying and forgetting our spychipped identification leaves us out in the cold one time too many, we just might be persuaded that implanting an RFID tag into our flesh would be the solution. And that would be a *very* bad idea—as you'll see in Chapter Fourteen.

8

TALKING TRASH

▶▶ THE RFID TAGS YOU THROW AWAY SAY AS
MUCH AS THE ONES YOU KEEP ◀◀

Your garbage can is like a trap door that opens on to your most intimate secrets; what you toss away is, in many ways, just as revealing as what you keep.

—Chris Lydgate and Nick Budnick, journalists who made a protest rummage through the garbage of Portland government officials[1]

Look at someone's garbage "and you know what people eat, what they are drinking, if they smoke, if they have kids, animals. You can see the personality."

—Pascal Rostain, celebrity garbage snoop/artist[2]

It's garbage day in the upper-middle-class enclave of Possum Hollow, a subdivision in unincorporated Dripping Falls, Texas. The trusting folks here haven't seen much crime in the ten years since the first foundation was poured in late 1999—unless you count a few recent burglaries, but they figure that happens in even the best of neighborhoods.

The residents roll out their trashcans in anticipation of the sanitation truck that shows up fairly predictably every Wednesday afternoon. It's this predictability and suburban naiveté that makes John's business so easy—and so satisfying.

Every Wednesday morning just after the streets clear of commuters and babies are put in for their morning naps, John scans the three hundred garbage cans in Possum Hollow for "business information." Armed with a car-mounted

RFID reader, he drives by the giant plastic receptacles and downloads information about the contents of each can onto his trusty laptop.

Who would have guessed that endless strings of ninety-six-bit numbers flashing up on the screen would be so valuable? John makes what he characterizes as "a very nice living" matching up skimmed information with street addresses for his clients.

While his interest in what the ritzy neighborhood throws away is strictly entrepreneurial, his clients obviously consider the information very valuable. It's so valuable that his innocuous little side business will soon allow John to upgrade his transportation to something quite a bit more impressive than his aging datamobile.

INCREDIBLE, BUT ENTIRELY POSSIBLE

Tidbits about John's entrepreneurial garbage surfing are raising a few eyebrows, we know. You're a bit skeptical that garbage could speak to marketers, snoops, and even criminals through remnant RFID chips. But we'll prove to you that not only is it technically possible to remotely read people's RFID-tagged trash, but businesses have already begun thinking up ways to take advantage of the spychips they expect us to unthinkingly wheel to our curbs on trash day.

When confronted with the serious privacy downsides of RFID technology, proponents fire back with crafted responses to defuse the concerns. Of course, we've already shredded arguments like "even if we could track consumers, we wouldn't have any interest in doing so." But our job is not complete until we've demolished one of the industry's favorite fall-back positions: the idea that RFID applied to product packaging poses no threat to consumers. "You're protected. Just throw it away," they say.

EPCglobal, the organization managing the numbering scheme for the RFID Internet of Things, has gold-plated this "trash talk." In its *Guidelines on EPC for Consumer Products*, the organization says we don't have to worry about item-level EPCs (electronic product codes) on RFID tags because: "It is anticipated that for most products, the EPC tags would be part of disposable packaging or would be otherwise discardable."[3]

But not so fast. There are several reasons why throwing away spychipped product packaging won't be enough to protect our privacy:

1. RFID tags on packages could allow marketers and others to track shoppers in the store, long before they reach the checkout counter to disable or discard the tags. As we explained in Chapter Five, this type of in-store tracking is already being planned in great detail. The "discardability" of the tags after the fact offers no protection at all.

2. Throwing away a spychipped product package could give consumers a false sense of security if the company has hidden a second chip in the product itself for surreptitious tracking purposes.

3. Many products and their packages are inseparable, so you can't just throw the packaging away. For example, you aren't likely to toss out the milk or window cleaner containers until their contents are used—which could be several RFID transmissions away.

4. And now we come to the point of this chapter: In the RFID world, trash talks!

Would They Really Do That?

Yes, there really are plans for all the stuff you throw away in the RFID world, and they're pretty dirty. But before we get to that, you need to see that there are people who would want to go through the trouble of gathering information about discarded items like old shoes, Coke cans, cardboard boxes, and stationery.

Portland police officer Gina Hoesly found out the hard way that there are those willing to brave maggot-infested table scraps, smelly cat litter, dirty diapers, and even discarded feminine products to learn more about our lives. Based on a tip that Hoesly might be using drugs, her boss ordered her fellow cops to snatch her garbage from the curb and search through it for clues. They thought they hit paydirt when they discovered a used tampon. They had it sent to the state's crime lab, where her dried blood was tested for drugs, DNA, and even semen.

Invasive? You betcha.

The tampon turned out to be "clean," but the cops found other evidence in the trash that won them a search warrant for Hoesly's home. That search turned up evidence the police were eager to use against the twelve-year veteran of the police force, who was already controversial for dating high-profile guys like Godsmack's bass guitar player, ex-Portland Trailblazer Jerome Kersey, and her superior, Assistant Police Chief Andrew Kirkland. She also had a history of blowing the whistle on bureau misdeeds.

While evidence obtained from Hoesly's garbage was thrown out of court by the judge, the decision was based on Oregon's strict state constitution. Had Hoesly been in another state, had she used clear trash bags, had she not put a tight lid on her trash can, or had she otherwise not proven her intent to keep her trash private, the results could have been quite different.[4] It sounds unbelievable, but warrantless garbage searches have been ruled permissible by the United States Supreme Court, despite arguments that they violate the Fourth Amendment.[5]

ONE MAN'S TRASH IS ANOTHER MAN'S CASH

The first thing we would do was locate a suitable home. For example, Jack Nicholson's or Bruce Willis's. Next, we would find out when the garbage was being collected and grab it before the truck came round.

—Pascal Rostain[6]

For some, the murky legal status of garbage has made trash collecting big business. French photojournalist Pascal Rostain makes a living snapping pics of the discards of the rich and famous. After lining up his finds on a black velvet background, he photographs them and offers the results for sale at art galleries around the world.

Rostain hit on his bizarre livelihood after reading about a sociology professor who required his students to go through other people's garbage as a course assignment. Shortly thereafter, he was called upon to photograph famed French singer Serge Gainsbourg (affectionately known as "the dirty mouth of French pop") at the star's home. This access presented Rostain with a unique opportunity to nab the singer's garbage—so he seized his chance.

Fondling Gainsbourg's discarded cigarette packages and empty booze bottles gave Rostain a rush. "It was like the key to Gainsbourg," he is quoted as saying. "I felt as if I had a part of him in front of me."[7] Anxious to repeat the experience, Rostain enlisted a partner and flew to Los Angeles where he bought a map of the stars' homes, several very large suitcases, and many pairs of yellow plastic gloves.

In the fifteen years that have followed, celebrities like Madonna, Tom Cruise, Pamela Anderson, and Mel Gibson have all become familiar with Rostain's work the hard way—by being spotlighted in his invasive exhibits. The results can be embarrassing. The gossip and snickering surrounding a package of Depends adult diapers found in Larry King's trash were enough to compel the talk show host to issue a denial that the incontinence products belonged to him. "They must have been in someone else's garbage," he told the New York Post. "I've never heard of Depends. I wouldn't know what a Depends looks like."[8]

The clues in people's trash can impact them in unexpected ways. The world now knows, for example, that although Halle Berry is a highly paid spokeswoman for Revlon, she does not use the company's products herself. (In fact, not a single Revlon product appeared among the many empty containers of competing beauty products found in her trash.) The public can also contemplate the meaning of a pair of underwear tossed out by Antonio Banderas, wonder what Sharon Stone could possibly have done with thirteen cans of pear halves, and vicariously experience the dinner party that left a case of Chianti bottles and several Cuban cigar stubs in Arnold Schwarzenegger's Hefty bags.[9]

When Rostain's work was recently displayed at an upscale SoHo gallery, the cognoscenti plunked down upwards of six thousand dollars per framed photograph. Not surprisingly, many of the buyers were the very celebrities whose trash he had "profiled." This strikes us as an innovative form of privacy extortion. What multimillionaire celebrities wouldn't hand over a mere $6k to prevent others from hanging an intimate snapshot of their private lives on a living room wall?

BENJI THE BINMAN

While Rostain's garbage picking may be obnoxious, it is nothing compared to the work of Benjamin Pell, the British rummager known as "Benji the Binman." The Binman regularly sniffs out ripe news stories from garbage rotting on the curbs outside the corridors of power and is handsomely rewarded by British newspapers for his troubles, reportedly earning thousands of dollars per story he helps unearth.[10]

He's managed to humiliate the top of the political heap, digging up financial information on politicians, tycoons, and even a memo written to Tony Blair by his pollster, warning that the Prime Minister was growing "out of touch" with the public.[11] The Binman's other targets have included celebrities. A self-professed obsessive-compulsive, he is said to have amassed seventy-five full bags of trash from Elton John's home.

His finds have appeared in the *Times*, the *Sun*, the *Mirror*, the *Daily Mail*, *Independent*, *Guardian*, and *Sunday Telegraph* newspapers. The Manchester *Guardian* assures us that "readers of virtually every national newspaper will at some point have read a story inspired or informed by his rummagings."[12]

Government officials in the U.S. are not immune to the garbage browsing threat either. The press has snagged the garbage of Portland's mayor, police chief, and district attorney and itemized its contents for public perusal (over their strenuous but ultimately futile objections), and reporters once heisted the trash of former Secretary of State Henry Kissinger and President Ronald Reagan.

But in the U.S., the law typically comes down on the side of the snoop. Despite the intimate nature of garbage, in most jurisdictions, it is perfectly legal to go through someone else's curbside trash without permission or a search warrant. Garbage placed near the street for collection is generally considered "abandoned property," waiting for anyone with an interest to examine its contents.

TAKING THE DIRT OUT OF DUMPSTER DIVING

As picking up someone's trash from the curb is legal in most jurisdictions, it would seem the main deterrents to more widespread celebrity trash displays are

(a) having the guts to grab someone's trash in broad daylight, and (b) having the stomach to sift through moldy produce and used Kleenex for that occasional prize.

But what if there were a new way to get the dirt on people? Imagine how much easier life would be for the paparazzi, criminals, and snoops if they could scan garbage remotely, secretly, and silently, from the convenience and privacy of their own cars? RFID promises this kind of instant inside information with no public scene, no fuss, no mess.

Okay, sure, that might someday be possible, you say. But what does it matter to me? Unless I'm a celebrity or a politician, I've got nothing to worry about, right?

Think again.

BIG PLANS FOR YOUR GARBAGE

Noting that some 80 percent to 90 percent of junk mail catalogs go straight to the trash, inventor Robert Barritz (CEO of Isogon Corporation, an IBM supplier) has come up with a "method for determining if a publication has not been read," according to U.S. patent No. 6,600,419.[13] Inquiring junk mailers want to know what sort of ingrate throws away their valuable marketing materials without even opening them, and spychips are just the tool to provide the answer.

Barritz's little trick involves programming a unique serial number for each household into an RFID tag affixed to the catalog before it is mailed. The tag gets hooked up to a "mechanical seal or switch on the cover which completes the circuit to the RFID tag." Opening the magazine breaks the seal and kills the RFID tag. This means that any catalog that has *not* been read would "respond to an RFID scanner with its identifying information. Otherwise, it will not respond at all."

Just as hidden web bugs report back to marketers whether you've read their e-mail spam, this device would tell them who read their postal junk mail spam. The patent explains this invention will "benefit individual retailers and market research firms" since "data on an individual consumer would indicate whether or not to consider sending future publications."

The inventive genius behind this scheme has grand plans for collecting the data and sending it to a central computer system for processing. He envisions automatically scanning people's household garbage with RFID readers attached to sanitation and recycling trucks. He even describes how reader devices could be fitted into smaller vehicles, so marketers or entrepreneurs like our hypothetical John could download data by simply driving along the curbside on garbage day.[14]

What is the lesson of this patent? Be careful which marketing materials or catalogs you open in an RFID-laced world. Make the wrong choice and you could be inundated with junk mail.

▶▶   RFID-ENABLED GARBAGE CANS AND GARBAGE
TRUCKS ARE ALREADY HERE ◀◀

Back in 1995, residents of Santee, California, received color-coded waste cans fitted with RFID tags. The purpose of the RFID tags was to help garbage trucks fitted with RFID readers to distinguish between the bins in order to handle them appropriately and automatically.[15] Indala, a spin-off from IBM, patented an enhanced RFID reader system for waste pickup vehicles on October 15, 1996.[16] It's a way for the garbage trucks to determine which cans to dump into their trucks when there are competing sanitation companies servicing the same neighborhood.

PROTECTING YOURSELF WITH THE SHREDDER

If you're thinking you can protect yourself from snoops by shredding spy-chipped documents before trashing them, we have some bad news for you. While you could do a pretty good job destroying the paper, miniscule RFID tags could be small enough to escape the teeth of even the best cross-cut shredder.

Of course, a snoop would have to get pretty close to grab the data from the smallest tags, since their read range would only be a few inches. But someone like Benji the Binman, who doesn't mind sifting garbage through his fingers, wouldn't mind that. The U.S. Postal Service has threatened to someday embed such tiny tags into postage stamps and envelopes for tracking purposes, making it theoretically possible that snoops could learn who had sent you mail even after you had reduced it to a pile of confetti.

In addition to documents, there are plenty of other things in our trash that simply aren't candidates for the shredder, things like empty cereal boxes, old tennis shoes, and broken toys. And at least one major American corporation has its eye on a way to sort through those discards for a profit. It's BellSouth, parent company of Cingular Wireless and a self-described "Fortune 100 communications services company headquartered in Atlanta, GA, serving nearly 50 million local, long distance, Internet and wireless customers in the United States and 12 other countries."[17]

BellSouth's Intellectual Property Corporation has taken assignment on a patent application entitled SYSTEM AND METHOD FOR UTILIZING RF TAGS TO COLLECT DATA CONCERNING POST-CONSUMER RESOURCES.[18] While the application discusses the worthy goal of sorting recyclable materials, its creepy crux boils down to how the data contained in RFID tags on disposed items "may be collected, sorted, processed, and provided for sale."

Yes, you read that right, *they plan to sell data on our trash.* Of course. We should have known BellSouth was just another megacorporation waiting in the wings to swoop down on the data revealed once its fellow corporate cronies spy-chip the world.

> Information concerning a post-consumption item [i.e., a piece of trash] may be linked (by serial number, for example) with information concerning the pre-consumed [i.e., a brand new] item collected by other data collection systems. . . . By combining captured pre-consumer information with post-consumption information, the entire life cycle of an item may be tracked. This information may be useful to any number of entities, including retailers, manufacturers, distributors and the like.

BellSouth goes on to explain how such data could allow close tracking of product transport and use:

[T]he collected and processed data may be helpful to track consumer purchase versus use patterns. A pet owner who lives in Atlanta but has a cabin in the mountains may choose to purchase pet food in the mountains because pet food is less expensive there. . . . A recycling facility may find it useful to know where items dropped off at the recycling center were originally purchased. Grocery stores, pharmacies and retailers may find it useful to know how long it takes a particular item to go from being stocked on the shelf to being placed in a waste or recycling receptacle. . . . The information collected may be . . . valuable to particular industries.

Valuable, indeed. And while BellSouth might have looked to the likes of the "Binman" for inspiration, it plans to capture its trashy tidbits with a little more dignity, according to related patent application 20040133484.[19] There, the inventor describes an elaborate garbage "sorting apparatus" complete with picker arms and conveyor belts so that even the most disgusting RFID-tagged items can be read and cataloged in a database without sullying human hands. With such automation it's possible that everything hitting the dump could be recorded in excruciating detail, a fact that would likely prompt a new adage: "Be careful what you throw away."

What in the world was BellSouth—a phone company—thinking when they sought out a garbage patent that has nothing to do with communications? Considering that BellSouth has wiring extending into twelve million homes like so many tentacles, and that it connects many of those homes to the Internet through broadband, the company could do some interesting things with its patent, especially if it teamed forces with a company working to connect household objects to each other and beyond. Such a formidable team could track "product lifecycle" right *through* the home.

THE FUTURE OF TALKING TRASH

For this ambitious trash tracking scheme to become a reality, nearly all packages will have to contain RFID tags. At current prices of tens of cents a piece, this is unrealistic. But as we discussed in Chapter Two, companies have been working on ways to manufacture disposable, printed RFID chips for a lot less money—maybe even pennies a piece. Back in 2002, *RFID Journal* predicted that innovations in cheap RFID chips could usher in widespread deployment of RFID-equipped packaging by 2008, making "very real the possibility that within five years ordinary bags of potato chips and boxes of cereal will have RFID microchips and antennas printed on them during the commercial printing process—much the way bar codes are printed on most products today."[20]

In the RFID world, even ordinary paper could be read at a distance, thanks to packaging manufacturer Tapemark. They're promoting a way to embed chip-less RFID transponders made of nano-resistant fibers into packaging and product labels. These covert, miniature transponders will be invisible to the naked eye. The fibers are as small as five microns in width and one millimeter in length—that's just a fraction of the length of a baby's eyelash! The proprietary combination of these fibers embedded in paper could reportedly reflect a unique serial number up to five feet away, and they'll only cost pennies apiece.[21]

Or how about spychipping newspapers? Flint Ink, one of the largest ink manufacturers in the world, is developing RFID tags and antennas made of conductive ink that could be indistinguishable from "normal" printing ink. It just so happens the company already supplies newsprint ink to many of the newspapers in the United States, a fact that could help ensure a seamless transition to remotely trackable newspapers.

Editor & Publisher magazine gushed at the prospect, saying that such tags "could provide a series of snapshots of customer behavior over time and across space" to tell publishers and advertisers how people read newspapers. Publishing industry analyst Miles Groves agreed that remotely tracking newspapers would be of considerable interest to major retailers whose ads appear on their pages.[22]

Nor would tracking reading material be the exclusive purview of newspapers. Arbitron, the media measurement company, and Time Inc., the publisher of popular magazines like *Reader's Digest* and *Time*, hope to someday use spychips embedded in magazines to monitor readership, just as soon as the price comes down. They'd also like to track such crucial factoids as where in the house magazines are actually read. Thanks to RFID, they could literally follow reading material into the bathroom with you.[23]

We're guessing the publishing industry would even like to pursue those magazines to their final resting places, which will open up all kinds of possibilities for entrepreneurs like our garbage surfing John and the overseers of BellSouth's trash picker arms in dumps across America.

9

YES, THAT'S YOUR MEDICINE CABINET TALKING

Truth is like the sun. You can shut it out for a time, but it ain't goin' away.

—Elvis Presley[1]

Every August, Elvis Presley fans worldwide descend on Memphis, Tennessee, for "Elvis Week." The week, which has now expanded to a rollicking nine days, includes dozens of Elvis-themed events that culminate in a candlelight vigil on August 16 commemorating the King of Rock 'n' Roll's untimely death.[2]

The Memphis Regional Medical Center, "The Med," is particularly busy during Elvis week, as fans converge to tour the only medical institution in the world named in honor of their hero—The Elvis Presley Memorial Trauma Center.[3]

While the throng of diehard fans and Elvis impersonators are a fun distraction from the serious business at hand, they give generously to their idol's namesake trauma center—money that is badly needed. "The Med" is what is referred to as a "safety net" hospital, the only avenue many of the poor have to any kind of medical care.

Like most inner-city hospitals, "The Med" is overwhelmed, underfunded, and inadequately staffed to handle the mass of humanity that enters its doors for everything from life-threatening accidents to ear infections. An estimated seventy-five percent of the patients who seek attention there are minorities and either uninsured or on Medicaid.[4]

What to do? Hiring additional staff, adding additional equipment, or opening up clinics for preemptive non-emergency care could have all helped relieve the logistical burden. But someone had a better idea: supply chain management and RFID.

The Robert Woods Johnson Foundation, the same people who fund National Public Radio programming, bought into the idea and funded a three-month experiment in which staff affixed adhesive-backed microwave frequency RFID tags to the ankle of each trauma patient entering the emergency center. Then, patients were tracked by the unique identification numbers on the assigned active RFID tags via twenty-five readers deployed throughout the 250,000 square foot facility.[5]

The Federal Express Supply Chain Management Center at the University of Memphis lent its logistical expertise for the test, giving new meaning to the term "delivering healthcare." FedEx's reputation for rapidly dispatching and delivering cardboard envelopes and boxes apparently made their educational arm the "go to" source for helping process the inventory of underprivileged patients.

Would Elvis, who was renowned for giving generously to medical causes, have been happy that his namesake trauma center was tracking people like overnight packages? We envision him saying something like, "Don't be cruel. These are folks just like you and me. They're not some boxes."

A company called Alien Technology (we kid you not) landed in the middle of this bizarre humans-as-boxes experiment, supplying the controversial technology. It was an incredible opportunity for them to find willing subjects on which to test their 2.45 GHz battery-powered RFID tags that could reportedly spy a patient from up to thirty meters away (almost one hundred feet).[6] Not surprisingly, Alien Technology looked back on the event with fondness in its April 2004 press release, characterizing its tagging and tracking trial as "successful." To preempt any criticism, Alien assured the public that the patients "were informed of the purpose and nature of the trial, and were enthusiastic to participate."[7]

Fair enough, but we wonder just how legitimate consent is when it is obtained from a patient in an emergency trauma unit. ("Okay, okay, Doc, I'll sign whatever you want. Just sew my leg back on!") We further question how ethical it is to represent that underprivileged patients could exercise choice in the matter given their dire circumstances. After all, where else could they go?

Of course, it couldn't have hurt that "the system was unobtrusive to the patients," as Alien pointed out in its PR release.[8] Keeping the technology out of sight is a consistent factor in getting people to accept RFID, which may be why the industry keeps aiming for invisibility, while we consumers keep aiming for notice and transparency.

Likening people to inventory doesn't stop in the trauma unit. Hospital administrators, public health officials, HMOs, pharmaceutical companies, and retail drug stores all hope to use RFID to keep tabs on patient health, drug use, and even food consumption. Their plans extend far beyond the needy patients who have served as their guinea pigs to date.

A REAL FULL BODY SCAN

If proponents have their way, all patients will one day have a shadowy new partner on their healthcare teams: RFID. All the physical items patients come in contact with and the patients themselves will be tagged and tracked. Believe it or not, even bedpans might be fitted with RFID devices hooked up to sensors so the quality and quantity of bodily outputs can be monitored around the clock and broadcast wirelessly to nursing staff and computer databases.

A company called Precision Dynamics Corporation (PDC) is banking on this future. Among other things, PDC makes a spychipped patient wristband called the *Smart Band* which consolidates a person's medical history, account history, and insurance information all on one chip. Promoted as an "indication that hospitals and the medical device industry are working overtime to meet the needs of patients," the Smart Band will enable hospital employees to ID a sleeping patient from several feet away.[9]

If the idea of being scanned as part of a hospital network of inanimate and human objects makes you feel like just another piece of inventory, it's not surprising. RFID is, after all, inventory control technology, and it has a disquieting tendency to distort things to fit that worldview—whether they are beloved family members struggling for life or newborn babies bursting onto the scene. The computer sees only numbers.

Hang on, patients of the future. If you think *today's* hospitals are impersonal, just wait until the RFID guys get through with them.

MEDICAL ERRORS

The PDC website features a page titled "Why RFID Is Critical" with a bold heading purporting to explain "Why hospitals need to be on board with RFID."[10] Imagine you are a hospital administrator reading the following dire statistics:

> The importance of positive patient identification in reducing medical errors
> harshly hits home when considering between 44,000 and 98,000 patients die
> in the United States each year from medically related errors at a cost of $29
> billion. The leading cause of death due to medical errors is caused by patient
> misidentification, and specimen or medication misidentification.[11]
> —quote on PDC website attributed to "The Institute of Medicine Report"
>   by Dr. Mark Chassin and Dr. Lucian Leape

Sounds compelling, doesn't it? But there's a problem when the sun shines on this statistic. As Elvis would say, the truth ain't goin' away. The "Institute of Medicine Report" doesn't mention patient misidentification at all, much less does it assert that patient misidentification was "the leading cause of death due to medical errors."

Thinking we must be missing something, we contacted one of the study's authors, Dr. Lucian Leape of the Harvard School of Public Health. We sent Dr. Leape a copy of PDC's quote, figuring that if anyone would know the truth, he would. Within hours, Dr. Leape fired back this scorching reply:

This [PDC quote] is a complete misrepresentation. One might even say a lie, in that it clearly is intended to deceive. Misidentification is a problem, but not one that has ever been quantified to my knowledge and it was not a specific item in the Medical Practice Study.[12]

Far from being the leading cause of death, it turns out that patient misidentification is at most a minor problem. When we followed up with him, Dr. Leape estimated that misidentification accounts for fewer than 5 percent of medical errors overall.

The myth of widespread patient misidentification has been repeated so many times and in so many places, from major newspapers to medical industry publications, that it has taken on a life of its own.*

While it is true that there are 195,000 preventable deaths a year at the nation's hospitals due to medical errors, the majority are due to causes like postoperative infections, bed sores, pulmonary embolisms, and deep vein thromboses**—the sorts of conditions that occur when there is inadequate personal attention to patient needs. Strangely, adopting RFID systems that further depersonalize the caregiving relationship could actually make things worse. At the very least, it would divert money away from efforts that really *do* save lives.

According to Dr. Samantha Collier, one of the study's principal authors, "If we could focus our efforts on just four key areas—failure to rescue, bed sores, postoperative sepsis, and postoperative pulmonary embolism—and reduce

---

*PDC is not the only company using this faulty statistic to promote RFID. We have caught at least two other companies selling medical RFID devices with the same pitch. A company called InfoLogix sells "next-generation systems [that] promise to pair bar codes with tiny transmitters like those used by highway toll booths to read speed passes, letting nurses identify patients from across the room." The company's promise of "accurate ID, tracking, and processing" made us feel like we were looking at a cattle auction near a meatpacking plant, not a place where vibrant human beings work together for optimum health. The InfoLogix website rhetorically asks visitors, "Why do hospitals need to be on board with RFID?" They answer it with the industry's standard (erroneous) line that "the leading cause of death due to medical errors is caused by patient misidentification."

**"The PSIs [patient safety incidents] with the highest incident rates per 1,000 hospitalizations at risk were Failure to Rescue, Decubitus Ulcer, and Post-Operative Sepsis. These three patient safety incidents accounted for almost 60% of all patient safety incidents studied."[13]

these incidents by just 20 percent, we could save 39,000 people from dying every year."[14] Given the low rate of death from patient misidentification and the serious problems elsewhere, it would seem wiser to invest scarce hospital resources in addressing real medical errors, rather than pouring money down the RFID hole.

TAKE YOUR MEDICINE—IT'S GOOD FOR ... US!

The pharmaceutical industry is overjoyed at the idea of adding RFID to their bag of profit-enhancing tricks. They'd love to get right into your bathroom and monitor every pill you take to increase something called "prescription compliance" (i.e., filling and refilling every prescription your doctor writes and taking every pill on schedule). According to statistics, while three out of five doctor visits result in a prescription being issued, only 50 percent of those prescriptions are ever filled. Of those 50 percent that are filled, 30 percent are not refilled.[15] What's more, some patients do not follow a strict protocol and might skip doses.

Whatever else they might represent, those skipped doses and unfilled prescriptions also mean lost revenue to drug manufacturers and retail pharmacies. CVS Pharmacy, with over five thousand stores in thirty-six states, fills an estimated 11 percent of all U.S. prescriptions. It estimates that improving prescription compliance by just 2 to 3 percent would mean the company could fill an additional 6.8 to 10.2 million prescriptions per year—something they call "a real opportunity."[16]

CVS notes that "because volume is a key driver, as an industry we have to find new ways to deliver on the expected increase." How will they do that? Why not install prescription-monitoring devices in patients' homes to ensure that they are properly consuming and refilling their prescriptions?

Once again, consulting giant Accenture rushes to the rescue of big business with an outrageous contraption it hopes you will welcome into your home. Its patented "Online Medicine Cabinet" is rigged with a high-tech camera, RFID, and an Internet connection so it can communicate directly with your doctor, the pharmacy, and maybe even your HMO.[17] Skip a Prozac and you're bound to get a warning. Do it twice and, well, you get the picture.

Intel Seattle researchers have also smelled opportunity in prescription compliance and lighted on the idea of a prescription nanny they call the MedPad. This appliance is a "flexible, low-overhead ubiquitous system" that monitors the patient's prescription use courtesy of embedded spychips and sensors.

Each medicine bottle has an RFID tag affixed to it that is read by the MedPad's integrated RFID reader. The pad recognizes when medication is removed and put back. In addition, the system sensor detects the difference in the before-and-after weight of the medicine bottle to ascertain what quantity of medicine was removed from it, and, theoretically, taken by the patient.

One of the MedPad inventors clarifies that, "By 'medications,' we mean any packaged item taken for health reasons, including vitamins, aspirin, cold medications, glucose tablets, and so forth."[18] Yes, they want to track it all, and the MedPad would make it possible.

So what does the U.S. Food and Drug Administration (FDA) think about all this RFID monitoring? As you'll read in a later chapter, heads of federal agencies have been directed to support RFID wherever possible, so the FDA is acting like a lovesick schoolboy around the technology. Per their marching orders, the FDA is claiming that the technology will reduce theft and counterfeiting of prescription drugs.

In 2004, the agency issued guidelines strongly recommending that all prescription drugs in the supply chain be RFID-tagged by 2007.[19] It even placed a rush on the implementation of its guidelines "because of the importance of beginning RFID feasibility studies as quickly as possible." Acting FDA Commissioner Dr. Lester M. Crawford said, "We hope that other manufacturers, wholesalers, and retailers will follow this example by also becoming early adopters of RFID."[20]

Several drug companies immediately stepped up to the plate. Purdue Pharma, the makers of OxyContin, announced they would incorporate RFID tags into the labels of 100-tablet OxyContin bottles shipped to Wal-Mart pharmacies. Pfizer began tagging Viagra containers shipped and sold in the U.S. in 2006. But we can't help but wonder if a man would really want to broadcast his impotence problem by carrying spychipped Viagra that could be scanned across a dinner table.

RFID on the Inside

How about swallowing an RFID device? The makers of the Jonah temperature sensor think it's a terrific way to measure a subject's internal temperature. The purple vitamin-tablet-sized device consists of an array of miniature electronic equipment including an RFID tag, a battery, and a miniature thermometer.

Once ingested, Jonah, presumably named after the famous Biblical character who languished in the belly of a big fish, travels the entrails of the patient, measuring her internal temperature along the way and beaming it out to monitors as far as three feet away. According to promotional literature from the MiniMitter (get it, *mini emitter*?) company that makes Jonah, "normal passage time is 1 to 5 days." Fortunately, MiniMitter assures us the Jonah device is disposable, so there's no need to reclaim it after each use.[21]

▶▶ Is There a Chip in Your Choppers? ◀◀

Your next dental prosthesis could include an RFID spychip. Dentalax, a French startup company, is promoting an RFID tracking system for dental work like bridges and crowns. During the casting process, an RFID tag is embedded within the material. As the prosthesis moves through different manufacturing stages, information is written to the chip via input from a PC hooked up to a reader device. This data includes patient information. Reportedly, the system has been put on hold due to privacy concerns. Let's hope so, because it would be nearly impossible for the average dental patient to know whether someone had planted a minute chip in his bicuspids.[22] Will they one day add a microphone? It could be on the drawing board, as you'll see in Chapter Fourteen.

As we dig into the files of the U.S. Patent and Trademark Office, we find further interest in capturing very personal data about bodily functions, as revealed by a bizarre patent for an RFID-enabled tampon detection system. When saturated, this device can send a signal from within a woman's body to a nearby computer reader, alerting the computer that it is time for her to attend to her personal hygiene.[23]

Even Kimberly-Clark, maker of Huggies, wants to get into the body fluid detecting act and has applied for a patent on a device that could be a boon for the parenting-impaired mother: the diaper with an associated sensor and RFID tag. The assembly senses when the diaper gets wet and sends a message to a computer system, which then alerts mom or other caregiver to change the baby.

But Kimberly-Clark sees much more potential in its invention and suggests an excruciatingly lengthy list of embarrassing places to monitor body excretions. These include everything from bed linens to toilet bowls to vomit bags and even suppositories.[24]

If this is all a bit too much to take, we understand completely, and invite you to join us in putting a stop to such invasive nonsense.

10

THIS IS A STICKUP

> ▶▶ RFID CRIMES, SHOULD-BE CRIMES,
> AND JUST PLAIN SNOOPING ◀◀

If someone walking down the street sees something they like—say, an article of clothing on a passerby—they can immediately take action.

—Accenture[1]

You're sitting on the train reading the latest Tom Clancy novel when some sixth sense makes you look up. Two seats away, a man is peering down at the LED panel of his Palm Pilot PDA. He presses a few buttons, then aims it pointedly in your direction.

Your eyes meet, and you get the uneasy feeling this guy has been watching you for some time. His glance leaves your face and moves down to the CVS pharmacy bag next to you on the seat. He presses another button on his PDA, then peers intently at the screen as if to check the results. Finally, he smirks at you and looks away. You feel yourself growing hot with anger. Who is this creep and what is he doing scanning you and your purchases?

Could this scenario be in our future? Could a voyeur sitting across from you on the train really electronically frisk you in 2010?

Yes, at least if you ask one major industry player. Accenture (the company behind the talking plant and the medicine cabinet we described earlier) is so keen to make this a reality they've patented a way for ordinary people armed with PDAs to click away at colleagues, neighbors, friends, enemies, and even complete strangers in order to learn more about what they're wearing and carrying. They call their patented vision the Real-World Showroom. We call it the pervert's best friend. Here's how Accenture says it will work:

> With the Real-World Showroom, consumers have immediate access to a wireless, always-on shopping channel. The showroom is, quite literally, the everyday world. If someone walking down the street sees something they like—say, an article of clothing on a passerby—they can immediately take action.
>
> How? The Real-World Showroom responds to RFID tags embedded in the item. By pointing a PDA—one with a permanent wireless connection to the Internet—at the item, it can be called up on the screen. Users can instantly find out more about the item and even purchase it.[2]

Unfortunately, this is not the vision of some small startup company. Accenture is one of the world's largest global technology consulting firms, with 2004 revenues of nearly $14 billion.[3] It counts Hewlett-Packard, NationsBank, Clorox, and Allstate among its "selected list of clients that have agreed to be mentioned."[4] There are others that apparently don't want to be mentioned. Makes a person curious as to why.

But while Accenture gushes about how companies can profit by turning the whole world into a gigantic sales floor, anyone who has ever been the victim of a crime sees a very different picture. The same scenario that could give strangers a kind of x-ray vision of what you're wearing and carrying could also be turned to other purposes, giving added character to the term "function creep."

The stealthy, invisible nature of RFID could make it the perfect tool for busybodies, snoops, and criminals of all stripes. Of these, perhaps most worrisome are the stalkers and abusive partners, whose use of this technology could have tragic consequences.

STALKING AND DOMESTIC VIOLENCE

Cindy Southworth of the National Network to End Domestic Violence knows this all too well. A social worker by profession, Southworth crisscrosses the country empowering police and hotline and shelter workers to help abused women stay hidden from violent ex-husbands and boyfriends. Since many abusers have threatened to kill these women or harm their children, Southworth knows their safety depends on keeping information about their locations and activities as private as possible.

With concern in her voice, Southworth can tell you how stalkers have begun turning to high-tech devices to get around their victims' efforts to stay hidden. Among the new technologies she fears could be abused: RFID.

"My concern is RFID in combination with databases," she says. "If a woman buys a sweater and its RFID tag number is cross-referenced with her credit cards in a retail database, her abuser could use that information to track her." Far from seeing the benefits of a "real world showroom," Southworth worries about its tracking potential because, as she puts it, "abusers go to great lengths to find their victims."[5]

Keeping RFID tags off of the things we wear and carry could be an important safety issue for a surprising number of people. More than a million women and nearly half a million men are stalked every year in the U.S.[6] Disturbingly, nearly half of stalking episodes escalate into violence.[7]

The U.S. Justice Department warns that "stalkers are, by their very nature, obsessive and dangerous" and should always be considered capable of killing their victims. In addition, they commonly use "advanced technology such as global positioning systems, wireless remote cameras, and invasive computer programs" to gather information about their targets in order to intimidate them or share their information with others.[8]

Though Accenture's Real World Showroom does not yet exist, Cindy

Southworth believes it's just a matter of time before stalkers and violent domestic partners tune into RFID's surveillance potential for their own purposes.

Based on recent high-profile cases in which stalkers have been caught dipping into the technology cookie jar, we are inclined to agree.

Consider Ara Gabrielyan of Southern California, serving sixteen months in state prison for hiding a GPS tracking device on his ex-girlfriend's car to monitor her every move. In addition to calling his ex between thirty and one hundred times per day, Gabrielyan took to arranging "chance encounters" at places like the airport and even her brother's grave. A police spokesperson explained that stalking his victim had become "an obsession, an obsession to the point where 24-hours a day he had to know where she was, what she did, who she met and how she carried out her daily routine."[9]

Someone this obsessed would jump at the opportunity to scan the contents of his victim's car, purse, or suitcase at every opportunity. Spychips would enable him to conveniently paw through her belongings at a distance, minimizing the risk of getting caught. To top it off, Gabrielyan could have avoided the prison term, since there would have been no need to plant spychips on his victim or her possessions. Corporate America would have thoughtfully done it for him.

Or how about Erik Reynolds, a thirty-three-year-old ex-felon, who changed his social security number to conceal his criminal past and found work as a deliveryman for FreshDirect, a company that provides groceries to upscale Manhattan neighborhoods. His job gave him up-close and personal exposure to women, including knowledge of where they lived and what products they bought. He used this information to terrorize six women with threatening and obscene phone calls, allegedly telling one victim (who was eight months pregnant at the time), "I'm going to come up and rape you. I know you are home."[10] Upon his arrest, Reynolds admitted to using customer information from order forms to make the calls.

Imagine someone like Reynolds further assisted in his demented pursuits by an RFID tracking system that could keep tabs on his victims through the unique ID numbers in the very things he delivered to them. He could scan the women's purchases in advance, note the numbers associated with them, and then use that information to remotely stalk them from the privacy of his own home computer.

A prototype of such a system, called RF Tracker, has already been developed.

▶▶ **RF Tracker: Using RFID to Stalk** ◀◀

To illustrate the stalking potential of RFID, legal expert David Sorkin, a professor at the John Marshall Law School, has created a hair-raising website showing how easily databases could be set up to "track RFID tags and their owners around the world." Sorkin's frightening vision of the future, online at RFTracker.com, emulates a website created by shadowy figures "based in an undisclosed location, in a jurisdiction that does not impose government censorship on the collection and use of personal information." (Hey, wait, that sounds kind of like the USA!)

The site concept is very simple. Visitors can type a tag number into a field and click "match" to see if the name of the person who owns the item has been entered into the database. It works the other way around, too—a visitor can type in an individual's name to download a list of tags known to be associated with that individual. But this is just background. The centerpiece of the website is its tag-sighting feature.

As Sorkin envisions it, a network of rogue scanners would capture tag numbers from random passersby and upload them along with the time, date, and location of the sightings. If a stalker saw one of his victim's targeted tags, he could phone her up with specific information about her activities, thus ratcheting up the terror. ("Did you have a nice time at the mall yesterday?")

Where would these rogue scanners come from? Sorkin explains how such a site might recruit data collectors: "We're always looking to improve our database. If you operate an RFID reader (whether you're a merchant, a manufacturer, or just a hobbyist), please contact us to discuss arrangements for transmitting your data to us. In exchange, we'll be happy to provide you with enhanced access to our databases . . . [or] we may even pay you outright."[11]

Voyeurs

Criminal voyeurs are quick to capitalize on the latest technological gadgets, so we expect them to be among the first to glom onto spychip-based surveillance systems.

One of the most popular—and disturbing—ways to spy on people today is through dime-sized video cameras that allow criminal voyeurs to capture images remotely. Already, such cameras have been found hidden in hotel bathrooms, department store dressing rooms, and middle school locker rooms. A Florida Fire Department employee even found one positioned under her desk, carefully angled to look up her skirt while she worked.[12]

But modern-day Peeping Toms have one vexing problem, as a camera found hidden in the women's shower on a U.S. naval command ship illustrates: its battery was dead.[13] Transmitting images with a spycam requires a lot of power, and the small size of the cameras means their batteries have a short lifespan. One technical workaround involves rigging the cameras with motion detectors so they only operate when someone is in view. RFID spychips could make the cameras even more efficient by activating them only when a specific target is in range as identified by an RFID tag he or she is wearing or carrying.

▶▶ TITILLATING TRACKING:
IT'S MY DUTY TO WATCH YOU, MA'AM ◀◀

Unfortunately, voyeurs are not limited to professional criminals and long-time lowlifes. Sometimes the voyeurs are the very people society charges with keeping others in line. Caesars Casino in Atlantic City was fined after security employees spent hours ogling scantily-clad female employees through the sophisticated casino camera system they were in charge of controlling.[14]

Police in England apparently do the same thing with the remote control joysticks and zoom features that they use to control their pervasive network of cameras, making it all too easy to take leisurely ganders at unsuspecting women.[15]

Unfortunately, the law has failed to keep pace with technologies voyeurs use to spy on people. In fact, as of this writing, it is still perfectly legal in many states to use a hidden camera to look up women's skirts (a practice known as "upskirting") and look down their blouses ("downblousing")—provided the spying occurs off federal property and in a public place.[16]

This means that any time women work, shop, sit down at a restaurant, or simply walk through a room that's accessible to the public, they could be fair game for voyeurs and perverts with nothing but their imaginations to limit the data they can collect on us. As far as we know, there is absolutely nothing in the law preventing these people from fully exploiting RFID to spy on us, too. The technology is so new that it's unclear what laws would even pertain to it.

How might perverts abuse spychips? For one, they could scan passing women to determine what they were wearing under their clothes. Apparently oblivious to this possibility, garment manufacturers have been eager to spychip women's bras and underwear. But the Benetton boycott demonstrated the voyeur factor was not lost on the world's women.

▶▶ HIGH-TECH MALL VOYEURS ◀◀

RFID proponents often defend spychips by telling the public that all they contain is a number. While in most cases that is true, a number can be very telling. The proposed numbering system for RFID chips, called the Electronic Product Code, or EPC, has a defined pattern that can reveal intimate details to anyone with access to the right database.

To illustrate how a simple number can be invasive, let's follow Sarah as she window shops in the local mall on her lunch hour in the year 2010. We join her as she approaches a table of young tech-savvy business guys drinking coffee at a table outside of Starbucks. What she doesn't know is that these guys are packing a reader device able to scan the RFID chips inserted into the seams of Victoria's Secret underwear.

The guys are peering intently at the screen of a laptop computer that's been loaded with the Electronic Product Code numbers corresponding to items in the lingerie company's most recent catalog. They downloaded it from a website that sells spy cams, picture-taking binoculars, and other vehicles for "wildlife observation" and titillation. They've whiled away many lunch hours placing bets on which of the passing shoppers is wearing those sexy undies.

As Sarah nears the table, the guys figure there's little chance this middle aged mommy-type is into fancy underwear. She breezes past their RFID reader, then looks back, prompted by the ruckus at their table. She shrugs off their spontaneous high-fiving, "can you believe its," and laughter as the kind of thing guys do when they relive the final points of a sporting event. She's right, but clueless that her Victoria's Secret red brassiere, size 38D, and the matching red-lace bikini panties her husband bought her for Valentine's Day did the scoring.

PICKPOCKETS, MUGGERS, AND THIEVES

In perhaps its most obvious criminal application, RFID would give thieves a huge advantage to spy out valuables and identify easy marks. Armed with hand-held reader devices (or perhaps Nokia RFID-reader cell phones currently under development), would-be muggers could lurk in alleys to scan passersby or station themselves at mall exits to preview the contents of shoppers' bags, waiting to pounce on anyone carrying an expensive video device or a Rolex watch.

Strangers on the airplane could remotely rifle through your carry-on bag without having to touch it. That quiet guy slouched in the corner of the subway—is he playing a handheld video game or peering into your backpack? Gangs of thieves could scan the contents of parked cars right through their window glass.

In addition, the workers you invite into your home or office to perform services like carpet cleaning, painting, and plumbing could use handheld RFID readers to "case the joint" for a robbery.

Ironically, some in law enforcement have joined with manufacturers and retailers to promote RFID tagging in spite of its threat to citizen safety. They have been lured by the promise that if items carry onboard RFID computer

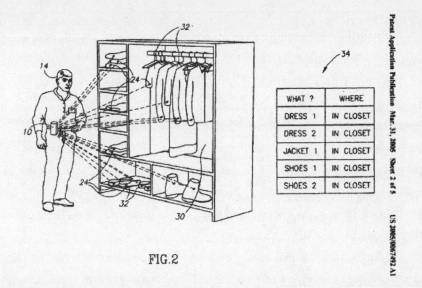

WHAT ?	WHERE
DRESS 1	IN CLOSET
DRESS 2	IN CLOSET
JACKET 1	IN CLOSET
SHOES 1	IN CLOSET
SHOES 2	IN CLOSET

Patent Application Publication Mar. 31, 2005 Sheet 2 of 5 US 2005/0067492 A1

FIG.2

IBM's vision of the RFID-enabled closet as revealed in U.S. patent application No. 20050667492.[17] The male figure is scanning items in the closet via the "personal index generator" hooked to his belt. While the person represented presumably owns the closet and the items in it, one can also envision him as the plumber, a snoopy guest, or even a technology-enhanced burglar.

chips with unique identification numbers, then stolen goods will be traceable and thieves won't be able to squirm out of convictions by explaining away their purloined possessions.

The Police Scientific Development Branch of the UK government has already spent millions in pursuit of this goal. In 2000, they enthusiastically launched their "Chipping of Goods Initiative" in concert with private companies including Unilever, Dell Computer, Woolworths, Argos, and Nokia. This chipping was "expected to assist investigators or police officers in identifying and recovering stolen merchandise, and be a powerful deterrent to would-be thieves."[18]

Thirty million dollars and several years later, they issued a more subdued assessment of RFID's antitheft potential: "For the UK, the Initiative has served to accelerate the awareness of tagging solutions and reinforce the association between achieving improved security and delivering increased business perform-ance. . . ."[19] Huh? If you read between the lines you will find that "powerful deterrent" is missing from the carefully crafted conclusion. And, of course, they didn't mention that a three-dollar box of Reynolds Wrap could foil their pricey system.

Even small-time shoplifters equipped with "booster bags" could swipe the tagged goods undetected since the RFID tags used in the Chipping of Goods Initiative could be blocked with ordinary aluminum foil. "Booster bags" are items designed by thieves to defeat shoplifting security systems. It's easy for thieves to make such bags. They just line a purse, backpack, or shopping bag with ordinary aluminum foil to block the communications between tagged goods and the security system. Creative thieves have even lined their underwear with foil—*ouch!*[20]

We doubt that spychips are going to stop retail crime sprees anytime soon, since criminals could so easily circumvent the RFID tag security system. Instead, regular consumers would be the ones to pay the price. This plan would require all of us to leave RFID tags live and functioning on our possessions forever. (Perhaps they haven't figured that out yet.) The only way to distinguish stolen items from legitimate items using RFID would be to link the tag numbers back to purchase records. This would require that each item be registered to a purchaser at the point of sale and the match fed into a massive database. Such a system would make law-abiding citizens vulnerable to surreptitious scanning, stalking, and a Pandora's Box of other crimes that are far worse than any theft problem RFID might solve.

HACKING, SABOTAGE, AND ESPIONAGE

There are plenty of ways that technologically sophisticated criminals can take advantage of vulnerabilities in RFID systems to commit crimes. These include eavesdropping, hacking, jamming, and more.

A team of graduate students from Johns Hopkins University demonstrated the hackability of RFID systems in early 2005 by defeating the Texas Instruments RFID "immobilizer" system. Millions of Nissans, Fords, and Toyotas rely on the system to prevent theft. The cars aren't supposed to start unless they sense a compatible RFID chip in the key. But the students worked out a method to siphon the code from the spychipped key and, with less than an hour's computing time, extract the code needed to hotwire the car.

Texas Instruments insists that such key cloning is unlikely, since a thief would have to get within twelve inches of a car key to pick up the information from its embedded chip. After all, no one except your valet, your mechanic, the people riding in the crowded elevator with you, the people waiting behind you in line, and the guy sitting next to you at the movies would ever get close enough to do that. (See a problem with TI's response?)

▶▶ IS IT TIME FOR RFID? ◀◀

(PHOTO: LIZ McINTYRE)

Speedpass rear window transmitter, Speedpass keyfob, and Timex Speedpass enabled watch. The RFID chip is normally hidden in the watchband right above the 12. Here, the plastic encapsulated chip is shown next to the watch.

Look Ma! No hands! Watchmaker Timex apparently thought 2002 was the right time for a hands-free version of the Mobil Speedpass, a key fob containing an active RFID tag that can be waved at Exxon and Mobil gas pumps to pay for fuel.

With great fanfare, they promoted a line of watches with Mobil Speedpass technology built into the watchband: "Speedpass™ and Timex® today released 4,000 first-edition timepieces that feature a miniature Speedpass radiofrequency transponder," they trumpeted. "The Speedpass-enabled Timex Watch looks and functions just like a regular watch, but the Dick-Tracy style device will enable customers to instantly pay for purchases at Exxon and Mobil service stations with just a wave of their wrists. Watch owners can even use it to purchase a Big Mac, fries and more at participating McDonald's restaurants in the Chicago and Northwest Indiana area." At one point, the Speedpass could even be used to pay for groceries at some New England area Stop & Shop grocery stores. They'd hoped this would be a "sneak preview" of a coming cashless society.[21]

But something happened. Mysteriously, the keyfob can no longer be used outside of Exxon and Mobil gas stations, and the Speedpass watch has been discontinued altogether. Even the "wave and go" option appears to be vanishing. To combat fraud, customers in certain areas must now enter a five-digit zip code to use the keyfob. Apparently, the system has been far too appealing to criminals.

Interestingly, the same defeated chip is used in the Mobil Speedpass contactless payment system.[22] So much for Mobil's promise that their Speedpass system is secure.

As for jamming, in the spring of 2004, Elgin Air Force Base gave an accidental demonstration of how easy it could be to interfere with radio frequency devices. While testing its new $5.5 million two-way radio system developed by Motorola, the base inadvertently jammed garage doors in the Florida panhandle communities of Niceville, Valparaiso, and Crestview, forcing puzzled homeowners to open their doors manually until the situation was resolved.[23]

If the technology exists to jam garage doors, imagine if terrorists used the same method to purposely jam systems monitoring warehouses full of goods.

And if they aren't hacking or jamming, they might be eavesdropping. At the Black Hat 2004 security conference in Las Vegas, IT consultant Lukas Grunwald showed attendees why gambling on RFID security is a bad idea. He demonstrated how anyone with a bit of computer savvy, some off-the-shelf equipment, and his RFDump software can read and potentially alter information on passive RFID chips from up to three feet away.[24]

Information about his RFDump software and a demonstration are available at RFDump.org.

TRACKING GOVERNMENT OFFICIALS

The United States House of Representatives is seeking proposals for a system to track lawmakers on Capitol Hill and pinpoint their locations in congressional buildings. They cite the need to ensure that everyone is present and accounted for in the event of an evacuation.[25] This is one of the worst ideas we've heard in a long time—and we've heard some bad ones!

Just as the perverts we discussed earlier in this chapter could use RFID tags to activate hidden surveillance cameras, so could spies. RFID tags associated with members of Congress could activate video and audio recording equipment whenever a selected lawmaker's RFID tag came within range. (Let's not forget the precedent set by U.S. Ambassador Averell Harriman who hung a bugged replica of the Great Seal of the United States in his embassy residence.) How convenient it would be for criminals and agents of enemy governments to have members of Congress broadcast their identities wherever they go.

One freedom-loving visionary has a whole lot more common sense than congressional security staffers. John Gilmore, cofounder of the Electronic Frontier Foundation and an early employee of Sun Microsystems, sees more malevolent uses for RFID. Here he explains how spychips could be used as homing devices for assassins equipped with "smart bombs":

In order to comply with government labeling mandates resulting from the huge Firestone tire recall, Michelin has announced that it plans to put RFID chips into every tire it sells to car makers (and eventually in all of its tires). Similar plans are afoot for other automotive and personal products.

Imagine being able to bury an explosive in a roadway that would only go off when a particular car drove over it. You could bury bombs like this months in advance, on every major or minor roadway. You could change the targeting whenever you liked (e.g., via driving a radio-equipped car over it and transmitting new instructions to it). You could give it a whole list of cars that it would explode for, or a set of cars and dates.

If you put such bombs throughout a metropolitan area, a car could drive through the area for months without triggering anything. But on an appointed day, each of the bombs in the surrounding area would know to go off when that same car passed. This would be possible without the responsible parties having to visit the sites later than days or weeks beforehand, making them hard to catch or deter.

Such explosives would be detectable by their radio emissions—RFID pings. But in a world where RFID pings are being transmitted by everything, including cell phones and door frames and cash registers and ATM machines and cameras and cars and computers and palmtops and parking meters and cop cars, you won't even notice. Places with "congestion pricing" like central London, or any toll road anywhere, will already have plenty of active RFID readers buried in the roadway.

Welcome to automated personal death. Courtesy of RFID and leading shortsighted global corporations, with government encouragement.[26]

Gilmore shouldn't be the only one worried.

Spychipped Passports and Terrorism

In another less-than-brilliant security move, the U.S. State Department began embedding RFID chips in passports in 2006. It's one thing if Congress wants to tag itself with homing beacons. It's quite another when they mandate that we ordinary citizens do the same. By the time this book is printed, thousands, if not tens of thousands, of chipped passports will have been issued, if all proceeds according to the government's plan.

The chip embedded in each passport contains the name, nationality, date of birth, and digitized photograph of each traveler.[27] Initially, the State

Department announced that the data on the chip would not be encrypted, giving rise to vocal criticism of the program.[28]

"This is like putting an invisible bull's-eye on Americans that can be seen only by the terrorists," said Barry Steinhardt, the director of the ACLU Technology and Liberty Program. "If there's any nation in the world at the moment that could do without such a device, it is the United States."[29]

Bruce Schneier, noted security expert and author of several books on digital security and cryptography, agreed that this initiative could make Americans sitting ducks, and he was puzzled as to why the U.S. government would jeopardize its citizens: "The only reason I can think of [for putting remotely-readable RFID chips in passports] is the government wants surreptitious access. I'm running out of other explanations. I'd love to hear one."

The State Department received 2,335 comments about the introduction of the electronic passport—98.5 percent negative. Despite this overwhelming opposition, the American public was overridden. The agency did not back down from its spychipping plans, writing that "by October 2006, all U.S. passports, with the exception of a small number of emergency passports issued by U.S. embassies or consulates, will be electronic [RFID-tagged] passports." But it did bow to pressure and agree to add shielding material in the passport cover to help prevent skimming of information by unauthorized parties.

11

DOWNSHIFTING INTO SURVEILLANCE MODE

[A]fter a relatively short period of tracking a vehicle, it may be possible to predict "when someone is or is not at home; where they work, spend leisure time, go to church, and shop; what schools their children attend; where friends and associates live; whether they have been to see a doctor; and whether they attend political rallies."

—*The Privacy Bulletin*, 1990[1]

HOUSTON, WE HAVE A PROBLEM

If you're driving on a major toll road in Houston and get the feeling you're being followed, don't ignore the sixth sense that has you double-checking your rearview mirror. While nothing may be obvious as you scan the road, there's a good chance that you are being electromagnetically probed, loop-detected, filmed, and cataloged without your knowledge or consent.

Information about you and your fellow drivers is being invisibly transmitted to Houston TranStar, the central nervous system of roadway surveillance in America's fourth largest city. If you could tour the TranStar nerve center, you might think it was designed by the same architect that outfitted the NASA Johnson Space Center just a few miles down the road, since it resembles the famous command center that handled the Apollo missions.

There are the same theater-style rows of workstations and communications gear facing a wall full of screens showing highlights of the action. But instead of astronauts floating inside the space capsule and the velvet expanse of the cosmos, workers at TranStar see snapshots of commuting vehicles, sliced, diced, and served up in detail for perusal by state and county civil servants, cops, and others with a stake in roadway compliance.

Wide-angle shots are beamed to an Internet site where the curious can click up images from some of the over three hundred cameras to see for themselves what's happening on busy stretches of metropolitan area roadways, but the crew at TranStar can get real-time, up-close looks at nearly every inch of pavement. The cameras, stationed at one-mile intervals, can zoom in on a target from a half mile away and rotate nearly 360 degrees horizontally and 120 degrees vertically. These Internet-enabled eyes in the sky are perched like robotic owls on poles at intersections so streets can be canvassed in all directions.

But this is just the start. TranStar promises citizens that even more cameras are on the way.[2] The government is working overtime to ensure your safe passage—or at least to watch you closely, safe or not. Brace yourself, there's more.

Houston has rigged over two hundred miles of area freeways and one hundred miles of carpool lanes with something called "Automatic Vehicle Identification (AVI) stations." Placed every one to five miles along the road, these listening posts probe vehicles equipped with RFID-enabled toll transponders to determine the vehicles' travel speeds.[3]

That's right. The toll transponders that millions of motorists have affixed to their windshields are now being surreptitiously used for more than just paying tolls. While Houston's roadside cameras will probably just see your car as a blip on the screen (unless the operator decides to zero in on you), if you have a toll tag, you're unknowingly beaming a unique identification number to roadway RFID readers, and the government is watching. It's a mind-boggling example of how technology can be introduced for one purpose, then evolve in unexpected ways over time.

At last count, Houston had installed 232 such reader stations to probe the over one million toll tags contained in Houston-area vehicles.[4] Oh, celebrate the gaze of the state!

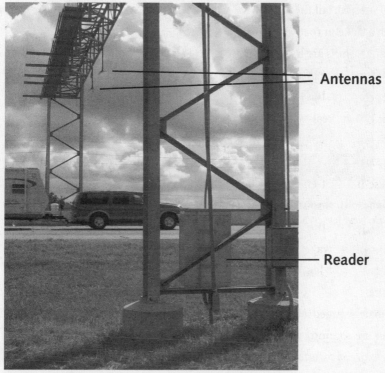

(Photo: Liz McIntyre)

This is a snapshot of one of Houston's many AVI stations. The antennas capture toll pass information that is sent to a central computer system for processing.

ELECTRONIC TOLL COLLECTION

Houston is not the only place in the country in pursuit of roadway omniscience. Other major cities across the United States and around the world are using RFID toll transponders that can personally identify drivers, or at least their cars. It's the price we pay for progress and greater efficiency, you might say with a wistful sigh.

(PHOTO: COURTESY OF RICHARD M. SMITH)

If found, please return to:
FAST LANE Service Center

CITIZENS BANK

FAST LANE

Matthew J. Amorello
Chairman, Massachusetts
Turnpike Authority

G3*021009

The Fast Lane toll transponder can be used in Massachusetts and on EZ-Pass toll roads in New York, New Jersey, Maryland, Pennsylvania, Delaware, and West Virginia.[5]

Fortunately, most toll roads still have a cash option that the privacy-conscious can use to preserve some anonymity, but those days may be numbered. To entice citizens to offer up their personal information to the roadway gods, transportation authorities have emphasized the convenience factor of toll transponders and even offered discounted toll rates. "Breeze through toll booths without the hassle of digging up correct change. Avoid the lines and save money, too!" they enthuse.

But when the soft sell doesn't work, some have taken to bullying tactics and arm-twisting coercion, like the State of Illinois. On January 1, 2005, holdouts who refused to use the Illinois I-Pass to drive on the Chicago-area tollways saw their fares double from the usual forty cents to a penalty "cash fare" of eighty cents.[6] This hefty surcharge amounts to two hundred dollars in additional fees per year for the average tollway commuter. Even so, some consider it a small price to pay to keep their travel records out of the hands of the state.

> ▶▶ PAY FASTER, EAT MORE! ◀◀

Here's a bad idea for people watching both their bottom lines and their waistlines. TransCore, the company that makes toll transponders for the North Dallas Tollway, is promoting the devices as a way to pay for drive-through fast food meals.

Already, five McDonalds in the Dallas area have arranged to accept the toll tags as a way to speed up the payment process. Why? They're banking on the RFID-based system to boost sales since studies have shown that customers who pay with toll tags spend an average of 38 percent more on food than those who pay by other methods.[7]

The last thing we need is a snoopy, high-tech way to encourage more Big Mac consumption.

Toll Records Are Telling

Authorities are doing all they can to get us to plant active RFID tags in our own cars, but they aren't keen to tell us the privacy dangers they pose. Those mini on-board spies can paint a detailed picture of our travels, ID our cars at a distance, and even keep track of a single individual's comings and goings. Toll tag records are already being used in unexpected ways, as one Illinois child custody battle illustrates. To establish that his client's wife wasn't spending enough time with the kids, a divorce attorney subpoenaed her I-Pass toll records to show a pattern of working late. (Can't be home and on the road at the same time!)[8]

Mandatory Spychipping of U.S. Cars

We've given you the gentle introduction. Now we're about to scare the bejeebers out of you. If the Federal Highway Administration has its way, in the future every car manufactured for American consumers will be spychipped before it rolls off the assembly line. This on-board spychip will be accompanied by a global positioning transceiver that can pinpoint the car by satellite and an 802.11 wireless device that can upload real-time location and vehicle data every time the car passes a roadside "hot spot."[9]

It sounds incredible, but that's the plan for Vehicle Infrastructure Integration (VII), an initiative being touted as a cutting-edge safety panacea slated for implementation within the next five years. With this system in place, cars will communicate both with each other and with a central system to avert collisions and prevent erratic roadway behavior.

You may have heard the one about the country bumpkin who got a flashy new car with the latest gadgets. Too lazy to drive, he turned on the cruise control and climbed into the back seat for a nap. In the VII world, he would not only survive with his car intact, but he would probably be blitzed with ads for hotels in his price range offering a more comfortable snooze and a free continental breakfast.

That's because there's a little-mentioned marketing catch to the VII initiative. (You knew there had to be one.) As cars pass hotspots, they would silently transmit personally identifying data to traffic management services and marketers. In return, marketers would respond with customized services like "streaming entertainment" and "interactive commerce."[10]

Access to this new inescapable in-car marketing channel will presumably be sold to paying clients as well as the companies putting the spying devices in the cars. This might explain why DaimlerChrysler, BMW, Ford, General Motors, Nissan, Toyota, and Volkswagen[11] are all supporting the plan in anticipation of all the new intelligence they can gather from consumers as they drive—consumers they desperately want to stay in touch with so they can sell them ongoing services.[12, 13]

Christopher Wilson, a VP with DaimlerChrysler, explains, "[O]nce we have this short-range network out there there's no sense in just limiting its use for [safety applications]." He adds, "We'd like to have a way to communicate with our vehicles when they're on the road. . . . A data link into the vehicle lets us do that."[14]

The revenue potential of VII has also captured the imagination of the bureaucrats. The all-knowing arrangement will let government entities identify exactly who is driving within their limits so they can charge a usage fee as well as keep tabs on the citizenry. This would allow cities like Boston to collect its

▶▶ THE CAR DEALER THAT WON'T GO AWAY ◀◀

We can count on our forward-thinking friends at Accenture to wheedle their way into any technological opportunity, and on-board car spy systems are no exception. They have already grasped the enormous marketing potential of "telematics"—the wireless extraction of information from vehicles as they drive past "hotspots."

Accenture wants to help car manufacturers siphon this information and use it as a way of "never saying goodbye to the customer." With Accenture's help, automakers may someday "stay connected with their products and customers long after the initial sale," since "telematics opens up the possibility of truly understanding customers' needs, wants and habits. Access to this kind of information provides a direct personalized marketing channel...."[15] As far as we know, there is no off-switch planned for this annoying idea.

Winston turned a switch and the voice sank somewhat, though the words were still distinguishable. The instrument (the telescreen, it was called) could be dimmed, but there was no way of shutting it off completely.

Oh, sorry—wrong quote. That one came from *1984*, not Accenture. (Oops!)

contemplated surcharge of up to five dollars per day for the privilege of crossing over from the suburbs[16] and enable the state of California to tax motorists by the mile, as proposed.[17]

We wish we could tell you the VII scheme was just a bad dream or a marketer's far-fetched fantasy, but unless we do something now, it's headed our way. It's slated for rollout between 2008 and 2010 through a gradual transition as consumers trade in their old cars for new spychipped models. They're laying the groundwork today. The U.S. Department of Transportation is "investing heavily" in VII applications,[18] and the Federal Communications Commission has already reserved a radio band for applications like VII,[19] so on-board spychips can beam data from cars traveling up to 120 miles per hour, from half a mile away.[20]

While we have some time to fight the invasive VII initiative, other RFID automotive applications could spring up overnight. The noose is already tightening with proposals for mandatory RFID-enabled license plates, RFID-enabled registrations, and even RFID-enabled inspection stickers that could also be used to monitor our travels.

Not surprisingly, the state of Texas tried to lead the charge with a proposed bill amending the Texas motor vehicle compliance code. It would have required special inspection stickers that "must contain a tamper-resistant transponder"

▶▶ A BOON TO BUSYBODIES ◀◀

Pastor Jim Norwood, the mayor of Kennedale, Texas, has an unusual hobby. He spends his spare time staking out porn shop parking lots, camera in hand, snapping photos of patrons' cars. He turns each photograph into a postcard, then mails it to the car's registered owner—along with an invitation to attend his church.

If we embrace RFID, Mayor Norwood's snooping ability could be enhanced beyond his wildest dreams. He could be spared the indignity of lurking in parking lots by installing an RFID reader at the entrance of every business he didn't like. Then he could automatically capture information from patrons' toll transponders, spychipped license plates, or vehicle inspection stickers.

Don't frequent porn shops? What about the supermarket?

Grocery giant Safeway was once caught copying the license plate numbers from one thousand cars parked in competitors' lots, then buying their owners' home addresses from the California DMV.[21] But someday the Safeways of the world may not need clipboards and cameras to capture our personal information. They could simply siphon our data by positioning RFID readers at strategic locations.

(a.k.a. spychip) to be affixed to motor vehicles. The transponder would carry the make, model, and vehicle identification number of the car to which it was affixed.[22]

The TransCore "eGO" tags lawmakers probably had in mind can be scanned remotely from over thirty feet away with a handheld or pole-mounted reader.[23] Alternatively, portable readers could be temporarily positioned along the road to monitor locations of interest.[24]

RFID in Driver's Licenses

National ID is a controversial topic. Its mere mention conjures images of jack-booted authorities demanding, "Your papers, please." A Gartner study in 2002 found "strong opposition" to the establishment of a national ID card due to "the potential for abuse."[25] But when it comes to driver's licenses, people are more accepting. So lawmakers seeking national ID went through the back door—our driver's licenses.

The Real ID Act signed into law in the spring of 2005 requires state-issued driver's licenses to contain a digital photograph, include anti-counterfeiting features, and be machine-readable. It also puts the Department of Homeland Security in charge of setting driver's license standards.[26] Given the cozy relationship between Homeland Security and the spychip industry, we predict we may soon hear a call for RFID in driver's licenses, too.

No Movement Unnoticed

Government-sponsored RFID transportation initiatives are promoted as ways to make us all more efficient and keep us safer. But no matter how they hype the benefits, downshifting into surveillance gear, a state where every move will be subject to approval and monitoring by Big Brother, could have some unexpected consequences as we'll see in the chapters that follow.

THE CHIPS
THAT WON'T DIE

It is desirable to continue utilizing the RFID tag as a data transponder, without destroying the tag or deleting its data memory, after an item containing the tag has been purchased at a point of sale.

—IBM patent application 200500733417 [1]

"How many seconds should we try?" Katherine asked as Liz looked on. She had loaded the first RFID tag into the old microwave oven retrieved from basement storage.

"I have no idea," Liz replied honestly. "I'd be conservative just in case."

We were both neophytes to tag-killing back in the summer of 2003. But in the interest of privacy, we owed it to CASPIAN members and the world at large to find out if spychips could be killed by zapping them in the microwave. Friends and supporters had recommended it as a possible solution to the looming RFID threat.

"Let's try fifteen seconds," Katherine said, confident that such a short time would meet Liz's "conservative" recommendation.

Beep, beep, beep, hummm. The microwaves rained down on the tag as we peered through the illuminated viewing window in anticipation of we knew not what.

"Oh, no!" we exclaimed simultaneously as the experiment poofed, arced, and flamed. Katherine frantically pushed stop to abort the test.

"How long was that?"

"Six seconds," Liz laughed, relieved that we didn't burn down the house. "Wow!"

Katherine removed the charred tag and wrote the cook time on the back of it. Then she reloaded, ignoring the smell of danger. "How long this time?"

"Three seconds at the *most*," Liz cautioned. "Let's try three."

Beep, beep, hummm. As the time end signal chimed, this tag was just on the verge of smoking as evidenced by the long brown char mark over the silicon chip. Katherine wrote the time on the back of the tag.

We fried a few more tags by zapping them for two to three seconds.

While we were successful in disabling passive tags with the microwave, we don't recommend you try this yourself. Not only is it dangerous, it would surely damage items in which the tags are embedded—and it could potentially harm the microwave itself. We've been hunting for a viable tag killer ever since—one that would not be burdensome, expensive, dangerous, or destructive. We regret to say that we haven't found one for general consumer use.

While drastic measures like crushing the silicon chip with a hammer and cutting the connection between the chip and its antenna kill spychips, the trick is knowing where tags are located and accessing them—not an easy task since spychips are easily hidden and removing them could damage the items. What's more, such low-tech measures are time-consuming and could pose other problems. We can envision a neighbor dropping by for tea, asking the kids where to find us. "Oh, they're out in the garage with their hammers beating on the new clothes they bought today to destroy any spychips."

We've heard many tag-killing suggestions over the years, including running the tag through the wash cycle, passing a magnet over it, or subjecting it to a

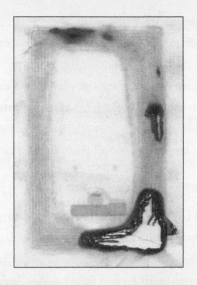

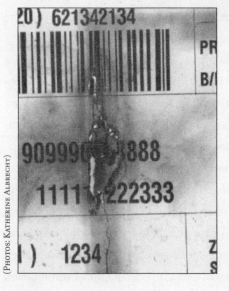

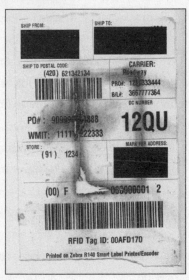

(PHOTOS: KATHERINE ALBRECHT)

Disabling RFID tags in the microwave works, but it's very dangerous. Don't try this at home.

VHS tape eraser. Unfortunately, these are *not* reliable fixes. Magnets and VHS tape erasers have no effect, and there are commercial laundry tags that can withstand high temperature washings and dryings.

Several inventors have contacted us with plans to develop "tag zappers," but so far, they are merely prototypes. We'll let you know on the Spychips website if we find a realistic tag-killing solution, but as of this writing, the best plan is to boycott stores that spychip their products so you don't have a problem in the first place.

SPELLING OUT THE ISSUES

To help alert the world to the looming RFID threat of spychipped products back in 2003, we helped author a document called the "Joint Position Statement on the Use of RFID on Consumer Goods." This document laid out the case against using RFID on individual products and was endorsed by over forty of the world's leading privacy and civil liberties organizations, including the ACLU, EPIC, the Electronic Frontier Foundation, the Privacy Rights Clearinghouse, and Privacy International.[2] There, we identify a number of RFID privacy fixes the industry has proposed, like killing RFID tags at the point of sale, having consumers tote around electronic devices to ward off spychip scans, and loading spychipped purchases in tag-killing kiosks. You can read the entire text of the position statement at our website: www.spychips.com/jointrfid_position_paper.html.

As we illustrate in the position statement, none of industry's proposed fixes offers a long-term solution to the problems RFID creates. In fact, many of these so-called solutions are merely smokescreens designed to appease consumers, while industry pushes forward with its plans unabated. As we've already seen, their first line of defense will be to assure you that you can "throw it away when you get home." If people refuse to take live spychips home, however, retailers may next offer to kill the tags at the checkstand when products are purchased, in much the same way they now disable anti-theft tags. But that's no good either. The only good solution is to keep products spychip-free from the outset. Here are some of the reasons why:

Killing tags after purchase does not address in-store tracking of consumers. To date, nearly all consumer privacy invasion associated with RFID on products has occurred in stores, long before consumers reached the checkout counter where chips could be killed. As we've already shown, the industry has big plans to use spychipped packages and products in conjunction with strategically placed readers to create the "retail zoo," where your every move is carefully observed and minutely recorded.

Tags can appear to be "killed" when they are really "asleep" and can be reactivated. Some RFID tags have a "dormant" or "sleep" state that could be set, making it appear to the average consumer that the tag had been killed. It would be possible for companies to claim they had killed a tag when in reality they had simply rendered it dormant. They could later reactivate and read such a "dormant" tag.

The tag-killing option could be easily halted by government directive. It would take very little for a security threat or a change in government policies to remove the tag-killing option. If we grow complacent and allow RFID tags to become ubiquitous in products, we could be setting the stage for an instant surveillance society down the road. All it would take is a single government directive to remove the safeguards (say, by making tag-killing illegal), and we'd be trapped in an inescapable surveillance world of our own making. Why hang the sword over our own necks in the first place?

Retailers might offer incentives or disincentives to consumers to encourage them to leave tags active. Customers who choose to kill tags might not enjoy the same benefits as other consumers. For example, they might not be eligible for sale prices, or they might not be allowed the same return policies.

The creation of two classes of customers. If killing tags requires some effort on the part of consumers, many will either be too busy or too trusting to bother. This would create two classes of consumers: those who "care enough" to kill the RFID tags in their products and those who don't. Being a member of either class could put shoppers at a disadvantage.

BLOCKER TAGS

What about using technical fixes to protect against spychip scans while we're in stores and other public places? Researchers at RSA Security have come up with an idea for something called a "blocker tag," an electronic device that would theoretically disrupt someone's ability to scan RFID tags by overwhelming their reader with irrelevant data. The idea is to embed the blocker tag in a shopping bag, purse, or watch that is carried or worn near tags with information consumers want blocked.[3] It's a great idea, but it has some problems:

Blocker tags are still theoretical. As of this writing, the blocker tag does not yet exist. Until a blocker tag is built and tested, there is no way to know how effective it could be or whether it could be defeated—either on purpose or because it stopped working naturally.

Encourages the widespread deployment of RFID tags. If people rely on them as a "cure" rather than pushing for "prevention," devices like the blocker tag might actually *encourage* the spread of spychips.

The blocker tag could be banned by government directive or store policy. Consumers could lose the right to use anti-RFID devices if the government someday decides that it's valuable to know what people are wearing or carrying. Retail stores might ban blocker tags as a "security measure." Once RFID tags and readers are ubiquitous in the environment, a change in policy could leave us all exposed and vulnerable to privacy invasion.

Adds a burden to consumers. Requiring consumers to deal with RFID tags themselves shifts the burden of privacy protection away from the manufacturers and retailers where it belongs and onto our shoulders. Do we really want to have to remember to carry a blocking device every time we go out if we don't want strangers remotely rummaging through our pockets?

You Can Trust Us . . . Really

All of this brings us back to the RFID industry's favorite solution . . . drumroll, please . . . *do nothing*. They want the tags live, so they can track your products for as long as you own them and even after you've thrown them away. But if you still need more reasons not to put such a powerful and dangerous tool in their hands, you need only look at their track record—and their patents.

Are these people you would trust with RFID?

13

ADAPT OR DIE

"Selling" the technology, the vision, or the consumer benefits exacerbates consumers' problems.... The best communication strategy appears to be positioning the technology simply as an improved barcode.

—Helen Duce, associate director Europe,

Auto-ID Center[1]

[I]n the case of EPC network there are currently no clear benefits [for consumers] by which to balance even the mildest negative.... The lack of clear benefits to consumers could present a problem in the "real world."

—Auto-ID Center Executive Briefing, 2003[2]

HITTING PAY DIRT

On a lark, Liz typed the word "confidential" into the website search engine, figuring it would flash back "no documents found." Instead, she gasped in amazement as the Auto-ID Center's website, the electronic heart of the consortium developing RFID, delivered over sixty confidential documents—and were they ever scorchers.

Clearly not meant for our eyes, the insider presentations and reports detailed plans to pacify consumers[3] and co-opt public officials.[4] There was even a private list of direct telephone numbers for some of the world's most powerful business executives—direct lines to the people who had funneled millions of dollars into the development of the spychip.

It was unbelievable. The very same people who had promised us that RFID data would be safe because "Internet security is very good"[5] had given us a compelling demonstration of exactly the opposite—and given us a close-up look at their dirty laundry in the process.

The RFID Industry Had a Problem

As we read the documents, it soon became clear the Auto-ID Center had problems that went well beyond their failure to secure their website. According to the studies they inadvertently made available to us, a whopping 78 *percent* of consumers surveyed reacted negatively to RFID on privacy grounds and "more than half claimed to be extremely or very concerned" about the technology.[6] Getting people to accept RFID was not going to be easy.

Did they take that as a sign to stop? Of course not. Rather than rethink their spychipping plans, they did what powerful corporations do: They threw money at the problem.

The Auto-ID Center hired the pricey public relations firm Fleishman-Hillard and set out to "develop best messages to pacify" consumers. Yes, pacify.[7] They came up with a plan to "identify potential consumer road blocks/fears, construct a proactive message framework to minimize negatives arising [and] assess consumer reaction if [the] press develop scare stories."[8] They conducted focus group research in North America, Europe, and Asia for insights to help them manipulate public opinion and prevent a consumer revolt.

And what did the people they surveyed say? Not surprisingly, the studies reported that their "biggest concern" was abuse. Consumers feared they could be tracked through their clothing, spied on by corporations and governments monitoring their purchases, and taken advantage of by thieves secretly frisking them. Their concerns read like a table of contents for this book:

- "I'd feel naked if people know what I'm wearing."
- "I could be tracked by the clothes I'm wearing."
- "Companies or the government will be able to monitor everything I buy and spy on me."

- "Someone could see everything I buy by reading my trash."
- "Muggers could know what is in my shopping bag or if I'm wearing a Rolex."
- "The technology will improve to allow people to read through the walls."[9]

Wow. Even at that early stage in 2002, consumers had an intuitive mistrust of RFID and the direction it would take. Unbeknownst to the participants, at the very time these interviews were conducted, the abusive scenarios they feared were being developed in corporate laboratories around the world. We know, because companies were eagerly applying for patents to corner the market on those invasive ideas.

The Strategy

The Auto-ID Center's advisors knew it would be tough to steamroll past the three-quarters of the population that would oppose their technology, so they proceeded on tiptoes and tried not to wake the sleeping giant. If their plans were discovered, they hoped consumers would feel hopeless and too "apathetic" to react (their exact quote was "on balance they are negative but apathetic").[10] Above all, they crossed their fingers that no privacy or consumer advocate would come along to shake things up.

"The best communication strategy appears to be positioning the technology simply as an improved barcode," they advised, noting that, ". . . discussing any benefits or using rational argument is largely ineffective and is perceived as 'spin.' Once consumers are concerned, they remain concerned, no matter what we tell them."[11]

In other words, keep it low-key, slip it in, and hope nobody notices until it's too late to stop.

Spinning: The Industry Sound Bites

At least, that was their strategy until we exposed their plans and began to fight back. Suddenly, the RFID industry could no longer rely on silence and tiptoeing to implement the spychip infrastructure. It was clear the public would never give corporations and the government the power to track every object on earth without a major battle. Unless the industry could find a way to spin RFID to make it seem more palatable, their plans were doomed. So spin they did.

The following are some of the "best arguments" they could come up with. See if you can spot the desperation in their numerous (and sometimes humorous) spins intended to drive RFID adoption:

Spin: It's Just an Improved Bar Code

Whenever you hear the term "improved bar code," you know you've been spun. As revealed in the Auto-ID Center's internal documents, that sound bite was identified as "the best communications strategy."[12]

But as we point out in Chapter Three, RFID is very different from a bar code. It contains a unique serial number, can be read through your clothes or backpack, and may cause health problems because of the electromagnetic radiation the reader devices emit.

Even the former director of the Auto-ID Center himself, Kevin Ashton, has commented on the stark difference, saying, "I think it is reasonable to compare RFID to the barcode, probably in the same way it was probably reasonable to compare the automobile to the horse. These are different technologies."[13]

Spin: It's Not RFID

In the trove of documents, we found the industry's PR advisor, Fleishman-Hillard, recommending a new name for the RFID tag to make it more palatable to consumers. Their suggestion? "Green tag."[14] We're guessing that was to conjure images of springtime, flowers, and caring for the environment.

Since we exposed that flash of Machiavellian brilliance, it's no longer an option, so the industry is looking to other possibilities for re-branding the technology. Who can blame them? The Procter & Gamble and Gillette RFID scandals have made the term "RFID" practically synonymous with spying and privacy abuses.

So a major re-branding effort is underway. The industry figures if they fool you with fancy verbal footwork, you won't figure out that what they've put in your shoe or your credit card or your passport is really RFID.

Tesco calls them "radio barcodes."

Marks & Spencer calls them "intelligent labels."

Wal-Mart calls them "electronic product codes."

The Auto-ID Center suggested calling them "green tags."

The Department of Homeland Security now wants to call them "contactless smart cards."[15]

Don't be duped. If a card, tag, or label contains an identification number or other information that can be read remotely via radio waves, it's almost certainly using RFID, no matter what they call it.

Spin: We're Only Doing This for Consumers

The industry has had several years to think up consumer benefits, and so far the best they've come up with is "faster checkout," "better product availability," and "improved recalls." While these are arguably positive things, they are certainly not critical enough to justify sacrificing our privacy and rendering our every belonging trackable.

There's a reason why the consumer benefits seem skimpy: This technology was designed by and for giant corporations, not consumers. Heck, we weren't even told they were doing it, much less consulted about the direction it should take.

The next time someone claims that spychips were developed to benefit consumers, remind them of the Auto-ID Center gem that opened the chapter: "[I]n the case of EPC network there are currently no clear benefits [for consumers] by which to balance even the mildest negative. . . . The lack of clear benefits to consumers could present a problem in the 'real world.'"[16]

Spin: The Tag Only Contains a Number

Refer to Chapter Three where we explain that every number could be associated with its own webpage and linked to unlimited amounts of information. If numbers are so safe, why don't we all wear T-shirts emblazoned with our Social Security numbers?

Spin: It's Got a Short Read Range

You don't need to be able to read tags from hundreds of feet away or from a satellite to invade people's privacy. In fact, sometimes a short read range is more powerful. For example, if you want to read an RFID tag in someone's shoe to determine exactly who is standing in a particular place, a short read range would be more effective than one that would pick up all the tags in the room.

Spin: It Would Cost a Fortune to Put RFID Readers Everywhere

Recall the reporter in Chapter Four whose sixty miles of travel could be tracked with just four reader devices. You don't need readers everywhere, as long as they're placed strategically.

Spin: Why Would Anyone Want to Collect All That Data?

What motivates Wal-Mart to collect and store twice as much data as the whole Internet? In part, it's because they can. With today's bargain basement prices for data storage, companies have little incentive to limit the amount of data they collect. They figure it could someday come in handy to refine store operations, run a marketing campaign, or even help government agents check up on a "person of interest."

Of course, no single human could sort through all those facts by hand. Instead, companies use a technique called "data mining" to extract the gems from the mountain of information they have collected. Think of it as looking up an article in a set of encyclopedias. You want the encyclopedia so when you need the information, you'll be able to find it. But that doesn't mean you're going to read it cover to cover.

Spin: Killing Tags at Point of Sale Solves the Privacy Problem and Putting RFID Tags on the Packaging Protects Consumers

We've already thoroughly debunked these myths, but there are a few more you haven't seen yet—so keep reading.

Spin: You Just Need to Be Educated

"Education" is the PR spinmeisters' word for propaganda designed to sell you on all the supposed benefits of RFID. The supposed "educational materials" we've seen fail to mention any of the technology's downsides, nor do they provide any meaningful information about how you can protect yourself.

Spin: We're Only Using It in Our Supply Chain

Many RFID proponents claim they only intend to use RFID in their supply chain, implying that consumers need not fear that it will ever impact them directly. But some people have an unusual definition of "supply chain."

Take Elizabeth Board, head of EPCglobal's Public Policy Steering Committee, for example. (Remember, EPCglobal is the organization that inherited the RFID "Internet of Things" from the Auto-ID Center.) Board recently explained that from her perspective, the supply chain extends all the way to the recycling center. So while consumers might see "supply chain" and think warehouses and distribution centers, apparently, some industry executives believe it includes your home, too.[17]

Spin: Spychips Will Keep Babies Safe in Hospitals

That's a diaper full of nonsense. Baby-snatching from healthcare facilities is actually extremely rare. According to a January 2003 report by the National Center for Missing & Exploited Children (NCMEC), out of approximately 4.2 million births per year at 3,500 birthing centers in the United States, abductions by non-family members are estimated at between 0 and 12 children per year. Of those, the baby is reunited with mom 95 percent of the time.[18]

Ironically, relying on RFID to prevent baby abductions could actually end up making a rare occurrence even more likely. Once hospital staffers rely on computer systems to track the human inventory in their care, they are likely to become less vigilant. According to the NCMEC, most abductions occur in larger, more impersonal hospitals.[19]

Spin: Once People Understand the Technology, They No Longer Fear It

Not so fast. We have spent a great deal of time studying the technology, and the more we learn, the more concerned we are. Even Ph.D. engineers who are experts on RFID have expressed concerns. Ari Juels, Ronald Rivest, and Michael Szydlo, three well-respected engineering and security experts who developed the "blocker tag" to mitigate RFID privacy risks, recently wrote, "The impending ubiquity of RFID tags . . . poses a potentially widespread threat to consumer privacy. . . . Researchers have recognized the RFID privacy problem for some time. . . ."[20]

Spin: RFID Will Save You Time and Money

Even if it saves you a minute here and a few cents there, the savings won't be earth-shattering, though the side effects will be. Procter & Gamble itself admits that the promise of time savings from RFID "may all appear to be a bit exaggerated and useless. At the end of the day it will save us only a few minutes."[21] Rather than universal cost savings, consumers are more likely to see variable prices. This could spell huge price increases for some shoppers—particularly the poor and those who enjoy bargain shopping.

Spin: We Have No Interest in the RFID Tag after Sale

Pure hogwash. The proof is in the large number of patent applications by major industry players that propose tracking live RFID tags after purchase. As we discussed, one patent application by IBM even proposes using RFID tags embedded in everyday objects to track persons of interest in public areas such as "shopping malls, airports, train stations, bus stations, elevators, trains, airplanes, restrooms, sports arenas, libraries, theaters, [and] museums."[22]

BACKED AGAINST THE WALL:

DESPERATE TIMES CALL FOR DESPERATE MEASURES

When it became clear that consumers weren't buying the stock sound bites, RFID proponents were forced to ratchet up their tactics. Soon spin was supplemented with "slurs" as the industry stepped into the ethical gray zone in pursuit of their goals.

On December 17, 2003, Katherine received an e-mail from the Grocery Manufacturers of America (GMA), a group that, according to its website, serves the interests of the food, beverage, and consumer products industry through efforts to "influence public policy" and "communicate industry positions to the media and the public." Their membership list reads like a who's who of the world's spychippers, and their support of the RFID initiative is well known. What is less well known are the tactics the organization uses to achieve its aims, including an attempted smear campaign.

According to e-mail evidence, a GMA employee e-mailed Katherine to request a copy of her bio, "for our sources." Katherine found the request unusual and responded by requesting further information. To her great surprise, the following day, she received a message that was clearly intended for someone else:

> I don't know what to tell this woman! "Well, actually we're trying to see if you have a juicy past that we could use against you."*

Concerned, Katherine requested an explanation from C. Manly Molpus, the CEO of GMA, and James Kilts, who in addition to being the chairman of GMA just happened to be the CEO of Gillette. Molpus issued an apology for the e-mail, explaining that the comments were made by an intern and were a "youthful indiscretion." He added, "Her request for a copy of your bio was simply part of a normal effort to obtain information about those who lead organizations with an interest in industry issues." However, GMA spokesman Richard Martin was later forced to admit that Katherine was the only person the organization had contacted.[23]

The incident quickly hit the headlines, and the press coverage was scathing. "Errant e-mail shames RFID backer," proclaimed *Wired News.*[24]

---

* Copies of the GMA e-mail exchanges are available at the Spychips website at www.spychips.com/press-releases/gma.html.

"Digital blunder exposes 'dirty trick' in RFID war," said CNET Network's online news site Silicon.com.[25] "Grocery manufacturers apologise [sic] to anti-RFID activist over slur," wrote Australia's *Sydney Morning Herald.*[26]

Bloggers had a field day, calling it a "smear campaign" and proclaiming that "RFID's intentions are as invasive and bad as the critics make them out to be." Activists everywhere shuddered.

Whether or not the e-mail was the start of an organized effort to discredit our efforts, we may never know. But even after the GMA backed down, other RFID players continued to sling mud in our direction. Here are a few of the more outrageous slurs we've had to deflect.

Slur: RFID Opponents Like Katherine Albrecht Are "Confused"

One of our favorite slurs comes from Dr. Cheryl Shearer, IBM's own global leader of business development for emerging markets. She characterized Katherine as "confused" and attempted to discredit her in an interview with ZDNet, a major online technology publication: "Katherine Albrecht has some sort of weird thing in her mind that helicopters might descend and follow you, I mean how low are these things going to fly?"[27]

For the record, we have never proposed that helicopters would be used to track RFID tags. After all, why use a helicopter when IBM has developed far more efficient, earthbound ways to track consumers? As detailed in their patent application, IDENTIFICATION AND TRACKING OF PERSONS USING RFID TAGGED ITEMS, IBM has developed an invasive "person tracking unit" that looks plenty effective.[28]

Perhaps Dr. Shearer needs help with her own apparent confusion. The "weird things" appear to be in the minds of Dr. Shearer's very own IBM colleagues who have developed such sophisticated ways to abuse RFID.

Slur: RFID Opponents Are "Fringe"

Derren Bibby, chief technologist at British IT strategy firm Noblestar and a zealous RFID proponent, called CASPIAN "some kind of fringe group in America" in his keynote address at the fall 2004 Enterprise Wireless

Technology show in London, England. He added, "These are the kind of people you need to watch out for."[29]

It's true that Bibby and his cohorts should be watching out for us, but it's not because of who we are, but what we know. Far from being fringe, CASPIAN represents a broad cross-section of the public, with members in over thirty countries from a variety of political, philosophical, and social viewpoints. They include business owners, homemakers, politicians, engineers, students, scientists, lawyers, factory workers, authors, and more. Katherine, CASPIAN's founder, is a former school teacher and holds a doctorate in education from Harvard. Liz is certified as a public accountant and worked for years as a bank examiner. (You certainly don't get to audit banks by being "fringe.") And, of course, we're both wives and moms.

Slur: Privacy Activists Are Alarmists

Frankly, there are times when an alarm desperately needs to be sounded. If that's what it takes to alert people to the danger headed our way, the alarmist label is one we don't mind wearing.

Slur: Only a "Vocal Minority" Is Concerned about RFID

Actually, the opposite is true, according to the industry's own studies. Only a small minority *isn't* concerned.

In October 2003, consulting firm CapGemini surveyed one thousand consumers and found that, in relation to RFID, "almost seven out of ten respondents said they were 'extremely concerned' about the use of consumer [RFID] data by a third party: 67% were concerned that they would be targeted with more direct marketing; and 65% were concerned about the ability to track consumers via their product purchases."[30] And that's a study conducted by an RFID *supporter*.

An Ohio-based marketing research firm called BIGResearch has reached similar conclusions. The company's March 2005 survey found that "the number of people concerned about the technology has stayed consistently around 65 percent from September [2004] through March [2005]," according to an industry analyst who works with BIGResearch.[31]

FROM SLURS TO SINS

With the majority of the world's citizens in opposition, the industry knows that "spin" and "slurs" alone might not be enough to rescue RFID from commercial failure. Spychip promoters have turned to even more underhanded methods, doing things we consider "sins" to promote their agenda.

Sin: Delay the Debate Until It's Too Late

RFID proponents like the GMA and EPCglobal are doing everything in their power to postpone a public discussion of the technology. They figure if they can get their infrastructure in place first, it's unlikely lawmakers and consumers will force them to dismantle it after the fact.

One delaying tactic is to portray the technology as "too new" or "not powerful enough" to warrant concern. Another strategy is to promise it's only being used in the supply chain, not on consumers. If they can delay long enough, they hope RFID will someday become "old news," raising little more than a polite yawn from cutting-edge journalists and the public. The time to discuss RFID is now.

Sin: Threaten Other Businesses—Adapt or Die

In July 2003, the Auto-ID Center organized an event to encourage major companies to get on board the spychip train. The invitation-only "CEO Summit" they sponsored at a posh Boston hotel was laden with a heavy subtext: Do it our way—or else. The not-so-subtle message was driven home by a book given to every senior executive in attendance. It was literally titled *Adapt or Die*.

Considering what happened next, it's hard to believe the book title was unintentional. Just months after the event, key Auto-ID Center sponsor Wal-Mart made good on the threat when it delivered a real live "adapt or die" message to its one hundred top suppliers. They would be forced to affix RFID tags to crates of products bound for Wal-Mart warehouses or risk losing access to the lucrative Wal-Mart market altogether. Failure to comply would mean losing Wal-Mart as a business partner, which, in turn, would mean financial ruin for most of these companies. So adapt they did, spending millions to comply with the mandate. The RFID train had left the station.

Sin: Co-opt Public Officials and Privacy Organizations

If you recall from Chapter Three, Katherine heard a senior executive comment about vocal opponents, "We should bring them in so that they . . . I don't want to use the word 'co-opt' but . . . we should make sure we deal with their [issues]." Well, it wasn't just talk. We've got hints of it in writing, too.

A confidential Auto-ID Center document identified "key government, regulatory, and interest group leaders" they hoped to "bring into the Center's 'inner circle.'"[32] These included:

- U.S. Senators Patrick Leahy and John McCain
- U.S. Representative John Dingell and W.J. "Billy" Tauzin
- FTC Bureau of Consumer Protection
- National Association of Attorneys General
- AARP (American Association of Retired Persons)
- AFL-CIO (A federation of fifty-eight American labor unions)

Many of these organizations and individuals have yet to weigh in on the RFID controversy, but we certainly hope they'll take the public's side when they do.

The Ultimate Sin: We'd Never Use RFID on People

Of course, spychippers vehemently deny they have plans to track people at all, never mind putting RFID devices directly on people. But in the next chapter, we'll share proof that RFID has already been used to monitor people against their will, and government agencies have discussed doing it again. Will you be next?

14

ARE YOU NEXT?

▶▶ NUMBERING, TRACKING, AND CONTROLLING
HUMANS WITH RFID ◀◀

And he causeth all, both small and great, rich and poor, free and bond, to receive a mark in their right hand, or in their foreheads. And that no man might buy or sell, save he that had the mark, or the name of the beast, or the number of his name.

—Revelation 13:16–17

That same scanner in a Wal-Mart that is used to bar-code your goods can be used to identify *you* . . .

—Scott Silverman, CEO of Applied Digital Solutions[1]

"**B**ang!" went the rivet gun. Ronald LaFortune* flinched as a black identification bracelet was clamped around his right wrist. Within minutes, he was a fingerprinted, photographed, and numbered subject of the military's Deployable Mass Population Identification and Tracking System.

His relief at leaving the whirring, buzzing, and riveting technology of the processing center tent was short-lived. As he queued obediently in the next line of his saga, the warm Caribbean sun glinted off the unending tumble-

---

*Ronald LaFortune is a fictional person whose account was based on our interviews with Haitian refugees who lived through internment at Guantanamo Bay during Operation Sea Signal, witnesses with firsthand knowledge of the situation, and written accounts by the military and industry reporters.

weeds of razor wire and M-16 gun stocks, hinting at the horrors to come. He took a deep breath and surveyed the guard towers dotting the perimeter of what he thought would be a retreat from the violence he fled in his homeland. Instead of the anticipated sweet smell of liberty, his nostrils were invaded by the interminable stench that he would endure for the many months of detention in the island internment camp.[2]

That fetid smell punctuated his mornings and wafted by as he swatted the malarial mosquitoes that drilled into him every night before he drifted off into a fitful sleep that was interrupted by the nightly wailing of children. Boredom, stress, and hopelessness weighed heavier and heavier on the sea of humans who were crammed into the squalid makeshift tent camps.[3]

Since there wasn't much constructive activity for the internees, apart from picking the occasional vermin out of their food rations, they were able to focus fully on all the impediments to feeling like full-fledged human beings. There were several conditions that vied in the game of most grievous. The parched, dusty landscape made the scarcity of water unbearable, and their portable toilets were clogged with a mixture of human excrement and worms. Rats ran rampant.[4] There wasn't even a private nook where they could weep about the rapes they had endured, the friends and family who had vanished without a trace, and the homes they were forced to abandon to save their lives.

And then there were those cursed black bracelets embedded with radio frequency identification chips. Each chip contained a unique nine-digit number, like a social security number, that linked to a central database.[5] The guards could wave a high-tech scanning device in read range of the chipped bracelet to determine facts about the wearer, like name, date of birth, and assigned camp.

Ronald recalls that the bracelets looked a lot like the watches kids get in McDonald's happy meals—except you couldn't take them off. A few of the refugees attempted to chew them off in desperation or cut them off with crude knives fashioned from metal scraps they scavenged.[6] But they were hauled off to a prison that made their miserable camp conditions seem livable.

So Ronald and the other refugees reluctantly accepted the bracelets as itchy reminders that they were at the mercy of their host country and the men in fatigues who rounded them up in what could best be described as a human cattle drive.

The refugees had more in common with cattle than would be noted by the casual observer. The company that made the disturbing black bracelets and the scanning devices that could read the numbers on the embedded Radio Frequency Identification chips was American Veterinary Identification Systems, Inc. (AVID), of Norco, California. The technology used to monitor and control the refugees was developed to keep tabs on farm animals like cows and laboratory animals like experimental rats.

Ronald was one of some fifty thousand Haitian and Cuban refugees who were numbered, indexed, tagged, and tracked by the United States military at Guantanamo Bay, Cuba, in 1994 as part of Operation Sea Signal.[7] Ironically, Sea Signal was a humanitarian mission to pick up the "boat people" fleeing the poverty and political repression of Haiti and Cuba. To the best of our knowledge, it was the first time in history that a mass of human beings was forcibly monitored and controlled with Radio Frequency Identification.

It probably will not be the last.

SPYCHIPPING THE VICTIMS

You might wonder how much you should worry about technology that was used on desperate people fleeing third-world countries. The answer: plenty.

We are all potentially just another 9/11 incident away from martial law and forced monitoring. An outbreak of smallpox, the release of a "dirty bomb," or the perceived need to control a particular swath of humanity might be seen as justification to round up targeted classes of humans, tag them, and track them with this technology.

How do we know we're just one emergency away from being rounded up like cattle? At a February 2000 conference, "Bioterrorism: Homeland Defense: The Next Steps," sponsored by the Rand corporation, government officials discussed how local, state, and federal officials should respond to a bioterror incident. San Marino, California, fire chief John Penido explained how he "anticipated working with the thirty-two municipal fire departments in Los Angeles County and local, state, and federal law enforcement agencies to formulate a response." Presumably, he and the other public officials in attendance envision tagging the occupants of a terror zone with some form of RFID. Referring to the "Deployable Mass-Population Identification & Tracking System," or DMPITS (the veterinary tracking system used on the refugees at Guantanamo Bay), he said, "DMPITS can be easily adapted for public health emergencies," adding, "to be effective, it must be initiated immediately and be capable of handling large numbers of patients."[8] As these statements were made prior to the September 11 terrorist attacks, we can only imagine how much these plans have advanced since then.

But it won't take a national emergency for people to become trackable if things proceed on their current course. The technology is already creeping into our lives in the form of removable chips and appears to be growing more sophisticated and invasive, progressing to chips injected into the flesh—perhaps even embedded deeply in the natural pockets of our internal organs, as you'll read later in this chapter.

If the ubiquitous networks of reader devices are deployed as planned and then supplemented with GPS transceivers, RFID chips in the things we wear and carry, and even our own bodies, could create a global, borderless human tracking system where anyone could be pinpointed on the earth in real time.

TRACKING PEOPLE AS INVENTORY

As we've explained in previous chapters, allowing our clothing, possessions, and the cards in our wallets to become spychip beacons will enable third parties to keep tabs on our whereabouts. RFID readers strategically installed at building entrances and exits could serve as "checkpoints" to create a log of people's movements. Such scans will make it possible to create detailed reports on where and how people spend their time and to make reasonably accurate guesses about whom they spend their time with.

At an industry conference in October 2004, Paul Heino of Sundex

Information Systems demonstrated this type of "people-as-inventory" tracking scenario. Promotional materials from the conference describe how attendees were rigged up with RFID tags and tracked—exactly like inventory:

> As a delegate you will have an RFID tag in your badge, and Paul Heino will briefly explain . . . how RFID technology can track the movement of delegates (as "products") by tracking their movements around the conference. This demonstration will illustrate the tremendous potential for greater efficiency through RFID-centric automation. Don't worry; there's NO risk to your privacy.[9]

While Heino's conference attendees were willing participants, it's obvious that such a system could be deployed covertly. In fact, that's exactly what was done with the conference badges worn by high-ranking government officials at an important European security event held in Geneva, Switzerland, at the end of 2003.[10, 11] Prime ministers, presidents, and other high-level officials from around the world were secretly tagged with RFID-enabled identification badges at the World Summit on the Information Society. Fortunately, the read range was short and data collection was apparently limited, but it illustrates how easy it is to slip tracking devices into seemingly innocent items. This incident shows that no one is immune—not even world leaders with highly trained security staffs. And, as they were unaware of the threat, there were no comforting assurances that their privacy was protected.

Human tracking is a hard product to sell to the public. When asked, few people say they are willing to voluntarily allow their own movements to be monitored. This may explain why most early tagging programs have involved groups who can't really say "no." These include the military, government employees, school children, and prisoners.

MANAGING THE MILITARY WITH RFID

The military has supported the development of the spychip infrastructure from the start. The Department of Defense was one of the earliest sponsors of the Auto-ID Center and gave the technology a huge boost when it announced a mandate in early 2004 that its suppliers would be required to affix RFID tags to shipments bound for DOD warehouses.[12] Many credit this move, along with the similar mandate by Wal-Mart, with starting the ball rolling on supply chain and inventory management applications for RFID.

Not surprisingly, the DOD is now looking beyond crates and pallets to use the technology to keep track of its most important inventory: soldiers. During a field trial in Iraq, wounded soldiers were fitted with RFID-enabled wristbands to track them from the battlefield through their treatment in a navy hospital. Handheld RFID readers equipped with a GPS module wrote information to the tags and recorded the location where each soldier was found. This information was updated throughout their treatment. Precision Dynamics Corporation, the company that supplied the wristbands, indicates on its website that the same technology has been used in Iraq to track enemy prisoners and refugees along with hundreds of wounded soldiers and airmen.[13]

You may recall Precision Dynamics from the earlier discussion about healthcare applications of RFID. They were the company we caught misquot-

(PHOTOS: COURTESY OF PACIFIC NORTHWEST NATIONAL LABORATORY)[14]

"Dog Tag" Reader/Writer.

RF-Linked Smart "Dog Tag."

ing medical studies and inflating the statistics about medical errors to promote their hospital tracking product.

But wristbands are just the beginning. Eventually, the military hopes to outfit each soldier with an RFID-enabled digital dog tag that could wirelessly transmit his or her name, rank, and serial number.[15] The men and women serving our country fear the next step may be a subdermal RFID implant. We've received letters from servicemen who are concerned at the prospect. Some have told us they would rather face a court martial than be chipped.

School Kids: The Youngest Captives

To the spychippers at Texas Instruments, the nation's fifty-five million school children[16] must look a lot like dollar signs. At least that seems to be the case since the company has apparently sighted inner city schools as a lucrative target market for its RFID card and reader systems. In 2003, TI landed a contract to use RFID to track students at the Enterprise Charter School in Buffalo, New York.[17]

Enterprise belongs to a network of sixty New York charter schools with an 85 percent minority enrollment and a high proportion of kids on public welfare. Seventy-five percent of kids attending New York charter schools are so poor they qualify for state-subsidized lunches.[18] As far as we've been able to determine, Enterprise was the first school in the U.S. to require mandatory RFID tags to attend school. Now the school's 460 kids are forced to wear Texas Instruments RFID tags around their necks that can be scanned from eighteen to twenty-four inches away. When scanned, the tags call up the kids' photos, dates of birth, and enrollment details as they enter the school building or go up to its second floor.

"I've been labeled the devil," says school director Gary Stillman, of criticism he's received over the mandatory program. But he's proud of the fact that none of the complaints have come from the kids' parents. "So far we haven't had one parent even question it. Not one. Because it's just so simplistic. We're just taking attendance," he says.[19] Of course, Stillman doesn't mention the fact that the school has a waiting list to get in, and students must reapply each year to continue their studies. Even had they heard of RFID and its dangers, how many of these low-income parents would rock the boat if it might mean losing their child's coveted spot at the exclusive school?

Then there's the question of cash-strapped New York schools, struggling to fund even basic necessities like teacher salaries and textbooks, spending tens of thousands of dollars to install this system. It hardly seems like a wise use of resources.

Another school district considering a Texas Instruments RFID tracking program is the Spring Independent School District just north of Houston, Texas. The Spring District wants RFID badges to monitor its twenty-eight thousand students as they get on and off school buses—supposedly, for the safety of the kids. Despite the fact that no child has ever been lost or abducted in the Spring District, students will be RFID tagged "just in case" (and at a considerable cost, too).[20] The District's foolhardy adoption of the spychips would be unsurprising, though, since Texas is the home of the company making the chips. What's more, spychip proponent Hewlett-Packard has one of its own seated on the Spring School District board of trustees (Kirby Bergstrom), who happens to be the chairperson of the board's finance committee, as well as a member of its technology committee.[21]

Kids report that rather than making them feel safer, the program makes them feel demeaned and insulted. One fifteen-year-old put it succinctly, saying that the tagging "makes me feel kind of like an animal." According to the *New York Times*, there have already been suggestions that subdermal RFID implants would be a more secure alternative to the badges, since kids can't lose or trade a microchip embedded in their flesh.[22]

LAB RATS IN THE ALL-KNOWING CLASSROOM

"We propose to target early childhood education as a testbed for our technologies . . ."

—UCLA researchers working on RFID applications[23]

RFID badges can be used for a lot more than attendance—just ask the UCLA researchers who developed the "Smart Kindergarten," a spychipped fishbowl for watching four- and five-year-old kids. They've created a chilling Orwellian classroom where every object—including the children—is RFID-tagged and continually monitored. Microphones and video cameras record kids' every move and statement. Spychipped hats equipped with RFID tags called iBadges keep track of where kids are looking. Nothing escapes the computerized watchers.

> The "iBadge" is equipped with sound, localization and temperature sensors but weighs just a few ounces and is not much larger than a quarter. To set up a test room in his lab, [a UCLA electrical engineering professor] has embedded tiny sensors into objects commonly used in a classroom and strategically placed miniature cameras and microphones around the space. Specially tailored "Gilligan Island"-style hats, which will eventually be worn by students, have been fitted with sensors to track speech and movement. The books, blocks and people will be interconnected to each other and to a database that can sift through all the information that the sensors gather.[24]

The researchers explain that the classroom toys, "in the form of objects familiar to children, will allow the environment to be instrumented with [sensor] devices in disguise." This will "enable applications that require unobtrusive capture of a child's actions (e.g., capturing what a child says when she is reading aloud)."

Why would they want to capture what a child says when reading aloud? And why would they need to use a disguise? To help the teachers manage the impossible task of "continuously listen[ing] to all conversations in the classroom." The plan is for every audio signal to be recorded, archived, and annotated for the teacher's later use. "[O]ur goal is to embed wireless microphones and speakers at strategic places, objects in the environment, and perhaps microphones even on kids [themselves] for localized speech and audio capture," researchers explain.[25]

If this is what researchers are working on in their laboratories, the spy-chipped school ID badges we've seen elsewhere may be just the thin end of the wedge. If parents get complacent about kid-tagging programs like those in New York and Texas, the next step for the insatiable watchers would be to implement the UCLA-pioneered classroom fishbowl. Then, they'd push for the workplace fishbowl. Then, the home fishbowl.

Fortunately, not all parents have been so docile when confronted with efforts to tag their children. In late 2004, Alien Technology (remember them from the hospital tagging trial?) linked up with a local company to tag K-8 school kids at the Brittan School in Sutter, California. The plan was for kids' spychipped name tags to communicate with a network of readers in class-room and bathroom doorways so school administrators could monitor their movements at all times.

Heavy-handed school officials distributed the name tags and told students to wear them—or else. Superintendent Earnie Graham made comments like, "[The badge] is just like a textbook, you have to have it. I'm charged with run-ning the school district and I get to make those kinds of rules."[26] Graham struck observers as the quintessential administrative bully—exactly the kind of guy you would *not* want having the power to monitor your child's bathroom visits. But Graham and his cronies got their comeuppance when Sutter parents mounted a tenacious protest campaign that garnered the attention of the national media and brought in privacy groups like the American Civil Liberties Union (ACLU), the Electronic Privacy Information Center (EPIC), and the Electronic Frontier Foundation (EFF). Within weeks, the invasive RFID pro-gram had been withdrawn, and a "just say no!" precedent had been set.[27]

TRACKING ON THE JOB

While vigilance saved the day in Sutter, other RFID applications have snuck into society beneath our radar screens, gaining a toehold before people understood their privacy-invading potential. Now entrenched, these RFID applications are harder to get rid of. One example is the millions of RFID-chipped employee ID and access badges used to track employees' movements.

Every time workers wave a plastic card at a wall-mounted reader to open a doorway or get on an elevator, they are giving out valuable personal data about their location and activities.

While it may occasionally occur to workers that their badges' can squeal on their movements, there's one place they probably don't expect to find an RFID reader: the bathroom. But a company called Woodward Laboratories has found a way to embed a tag reader into a product they call the "iHygiene Perfect Pump." It's a liquid soap dispenser that doubles as an employee badge reader and monitoring device.

To unsuspecting employees, the device appears to be a perfectly normal soap dispenser. But hidden within its sleek plastic exterior is an electronic spy that captures the ID badge number of the person standing at the sink and watches to see if the employee washes his or her hands. A report of each employee's bathroom hygiene practices is then "easily transmitted via the Internet" to enable "enterprise-wide hygiene compliance monitoring."[28]

Though few would disagree that employee handwashing rules are sensible and important, it's a big leap from there to concluding it's appropriate to hide RFID readers in the soap dispenser to watch people when they're in the bathroom. There are many useful rules in society that should be obeyed. But do we really want to set a precedent that it's okay to secretly monitor people any time there is a rule in place? That's dangerous logic. While the person being monitored today could well be someone you don't know, tomorrow the camera could be focused on you and your loved ones. If being the monitored is something you'd rather not have happen to you, then you need to take a stand today to prevent such technological invasions from creeping into your workplace.

Taken to its logical endpoint, the same reasoning used to justify hidden recording devices in soap dispensers could one day be used to justify monitoring you in your favorite restaurant, your automobile, or even your home. After all, there are numerous rules you shouldn't be breaking, and keeping an eye on you twenty-four hours a day would help ensure your complete compliance.

The handwashing surveillance system requires employees to wear RFID-enabled badges, but soon employees' actual uniforms could report on them instead. The nation's top two uniform rental companies, Cintas (which clothes workers at Starbucks, Disney, Sears, and Wal-Mart)[29] and Ameripride (with clients like Outback Steakhouse, 3M, and Chevrolet)[30] have quietly begun slipping spychips into employee uniforms to keep track of washing and rental logistics.

The tags come encased in sealed plastic disks that can withstand years of commercial laundering, yet still beam out their unique ID numbers whenever they come within range of a reader device. Uniform companies aren't rushing to tell workers about the spychips in their trackable clothing, but they're eager to sell employers on the embedded spychips. AmeriPride's website, for example, shouts "STATE of the ART TRACKING TECHNOLOGY" next to an animated, spinning RFID laundry tag featuring the company's eagle logo.[31] And as far back as 1997, Accenture (who else?) slipped early model RFID tags into the waistbands, shirttails, and collars of eighty thousand uniforms worn by Australian workers as part of its "Silent Commerce" initiative.[32]

·Though the spychips are promoted as a way to track uniforms, not employees, it doesn't take a radio engineer to see the potential for tracking the workers wearing them. Signals captured from an employee's uniform could be used to time restroom visits, monitor trips to the water cooler, or measure time spent at a desk. They could even be used to record who is spending time with whom—not only curbing office romances but potentially chilling whistle-blowing as well.

Close to thirty million American workers wear rented uniforms to work each day, including workers in retail, manufacturing, law enforcement, food service, healthcare, transportation—you name it. We foresee an outcry when these millions of workers all take a close look at their uniforms and find the hidden tracking devices. Like the parents in Sutter, California, we envision them banding together, speaking up, and putting a speedy end to the hidden tags.

Embedding an RFID Tag in Your Flesh

Up until now, everything we've discussed can be removed. If you don't want the school or the boss tracking you, you can simply take off your badge or demand a new uniform. But what if the RFID device was a part of you, embedded in your flesh? There now is such a device. Manufactured by a Florida company called Applied Digital Solutions, the "VeriChip" is a glass-encapsulated RFID tag that is injected into the flesh, typically in the triceps area, midway between the elbow and the shoulder. The device has already been implanted into millions of dogs and cats around the world, and now its developers want to put it into people.

While it is usually described as being "about the size of a grain of rice," the VeriChip actually measures 12 mm (.47 inches) long, making it a bit shorter than the diameter of a dime. That's a lot larger than any rice we've ever seen—and we both eat long grain rice. Perhaps the company uses the "rice grain" description to help soothe the fears of potential implantees who might be understandably nervous when they see the intimidating hypodermic injection apparatus coming at them. From what we can gather, it's a pretty painful process.

Eighteen government workers in Mexico experienced this firsthand in June of 2004, when former attorney general Rafael Macedo de la Concha spy-chipped himself and many of his employees as a way to secure access to a sensitive records room. Rather than use a key or a swipe card to get in, the chipped employees pass by an RFID reader portal that scans their VeriChip implants. If an employee's chip returns an authorized number, he or she is allowed to pass through the door.

We don't know how many staff members were asked to take the chip or what the penalty was for refusal. (While many press reports indicate that 160 employees were chipped, we contacted their press office directly to ferret out the real story and learned that the initial reports were greatly exaggerated.[33]) Presumably, employees who refused were reassigned to jobs that did not require access to the room. We're guessing they probably felt pressure to comply with the chipping.

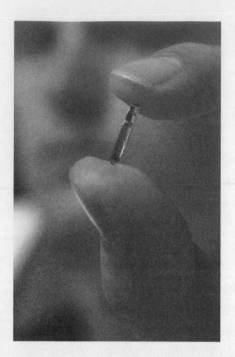

Photo of a VeriChip courtesy of Applied Digital Solutions.

The "Home Again" chip marketed by Schering-Plough for implantation in dogs and cats is essentially the same as the VeriChip RFID Implant for humans. Both are manufactured by subsidiaries of Applied Digital Solutions.

The whitish substance on the end of the chip is an anti-migration coating that encourages tissue growth so the chip doesn't move around inside of the animal—human, feline, or canine.

(Photo: Liz McIntyre)

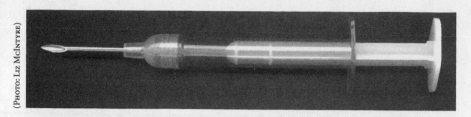

(PHOTO: LIZ MCINTYRE)

The cannula (injector device) to insert the chip into an animal. This is one honker of a tool. Applied Digital Solutions never responded to our requests for images of the "human" implantation device.

Not only is the program invasive (to say the least!), but it's foolish for the employees from a security standpoint, too, especially given Mexico's serious kidnapping problems. A criminal wanting access to a secure room or some-one's bank account will now be tempted to kidnap the person directly and remove their chip by force. The result could be quite gruesome. According to reports, at least one criminal gang in Mexico known as *Los Chips* assumed its wealthy kidnapping victims had a chip hidden somewhere in their bodies. They threatened their victims with violence if they wouldn't reveal the loca-tion of their implanted chips.[34]

Despite valiant efforts to market the VeriChip, including the creation of a rolling implantation clinic dubbed "The Chipmobile," and the dubious slogan "Get Chipped," clients haven't exactly beaten a path to the company's door. This may explain why they've now pinned their hopes on markets with fewer inhibitions: the drunk and the dead.

CHIPPING BAR HOPPERS

Applied Digital's first foray into the drunken market niche took them to the bar scene in Barcelona, Spain. There they found the Baja Beach Club, a nightclub catering to the under-twenty-five crowd, described as a cross between Hooter's and *Spring Break: The Movie.* To give you a feel for the place, one visitor described procuring "the services of one of the beach babes walking around. For eight euros, she will 'interact' with the [customer], her props being test-tube cocktails, whipped cream, her amazing body and lots of imagination!"[36]

▶▶ VeriChip: Invitation for
Kidnapping & Mutilation? ◀◀

It might seem convenient to use a part of your body as the key to valuable assets or sensitive areas. After all, it's always with you. But that's no guarantee you won't lose it. Just ask Malaysian accountant, K. Kumaran.

When a band of carjacking thugs grew tired of repeatedly forcing Kumaran to place his finger on the biometric touch pad to start his late-model Mercedes, they took a more direct approach. Using a machete, they hacked off his finger before leaving him naked and bleeding by the side of the road.[35]

While he may have lost his finger, had his car required a VeriChip rather than a fingerprint for activation, Kumaran could have lost his entire arm.

In March 2004, the nightclub staged a VIP chipping night featuring various B-grade Spanish movie stars. Several of them had achieved their fame by living in a house rigged with cameras and allowing their exploits to be televised on the *Gran Hermano* TV program, the Spanish equivalent of America's *Big Brother* series.

During the VeriChip promo night, a company representative in a white lab coat was on hand to inject the stars and any others foolish enough to purposely embed a microchip payment device in their flesh. The lure: the ability to breeze past the bouncer, access VIP areas of the club, and pay for drinks without bothering with cash or a credit card. As a scanner is passed within a few inches of the patron's spychipped flesh, the implanted RFID device transmits a unique ID number that can be linked with financial accounts and club membership information.

The bizarre event was such a novelty (and garnered so much media attention) that several other bars followed suit, including the Baja Beach Club in Rotterdam, Holland, a night spot called Bar Soba in Edinburgh, Scotland, and the Amika nightclub in Miami Beach, Florida.

Though a handful of people have undergone the procedure, it's still not considered mainstream, even at those clubs.

CHIPPING CORPSES

After minor successes chipping nightclub revelers, the RFID industry turned its marketing efforts to an even more acquiescent crowd: the dead. They were on hand with microchips to implant into the corpses of the victims after the devastating Southeast Asian tsunami,[37] and they're ready to help should research organizations like the University of California decide to embed RFID chips into cadavers and associated body parts. The school is considering microchips as one possible way to stop the illicit trafficking of human remains donated to their school in the wake of lawsuits by donor families.[38]

▶▶ PAYMENT IMPLANTS AND THE MARK OF THE BEAST ◀◀

The RFID implant device, known variously as the VeriChip, or VeriPay, sets off alarm bells for a lot of Christians. Many believe it may be the fulfilllment of a prophecy made back at the time of Christ. Revelation, the last book of the Bible, describes a time when all people will have to take a mark in order to buy or sell. Recall Revelation 13, which we quoted at the beginning of the chapter: "And he [the beast] causeth all, both small and great, rich and poor, free and bond, to receive a mark in their right hand, or in their foreheads. And that no man might buy or sell, save he that had the mark, or the name of the beast, or the number of his name. Here is wisdom. Let him that hath understanding count the number of the beast: for it is the number of a man; and his number is six hundred threescore and six," or six hundred and sixty-six.

We're not sure how the 666 part fits in, but an RFID implant linked to an electronic payment account—especially if it's placed in the hand or forehead—would bear an uncanny resemblance to the rest of this description. We know we sure won't be taking one.

The "Medical Device" That Does More Harm than Good

Perhaps hoping for some respectability, the folks at VeriChip's parent company, Applied Digital Solutions, have tried promoting their product as a lifesaving piece of medical equipment. In October 2004, the FDA approved the use of the VeriChip as a medical device to store a unique ID number linked to patient health information.[39] The idea is for Applied Digital Solutions to maintain subscribers' health records in a central database that could be made available to hospitals and paramedics equipped with handheld VeriChip readers.

Theoretically, in an emergency, medical personnel could quickly determine a patient's medical history, allergy status, etc., by waving the reader over the patient's arm, getting the number, and cross-referencing it in the central database. The problem is that few medical facilities have embraced the plan. They have a hard time seeing why people should switch from the Medic-Alert bracelet, a far more efficient, low-tech method for communicating serious allergy and medical information that has served the public well for over fifty years.

Plus, for a chip that is supposed to save lives, the VeriChip has a surprising number of medical downsides and risks associated with it. Some are so bad that, ironically, people with RFID implants may actually need to wear a Medic-Alert bracelet to tell medical personnel of the chip buried in their arm. Back when the company was boasting of its FDA approval, we poked around in their SEC filings and found a scorching letter from the FDA outlining a laundry list of serious risks associated with "an implantable radiofrequency transponder system." According to the FDA:

> The potential risks to health associated with the device are: adverse tissue reaction; migration of implanted transponder; compromised information security; failure of implanted transponder; failure of inserter; failure of electronic scanner; electromagnetic interference; electrical hazards; magnetic resonance imaging incompatibility; and needle stick.[40]

The line about magnetic resonance imaging, or MRI, would seem to be a real problem. Metal in the body, such as the antenna inside the VeriChip, has a disconcerting tendency to heat up and move through body tissues when exposed to the energy fields of an MRI. Theoretically, an implanted chip could overheat, causing the device to burn the patient or fail. It might also tunnel through someone's flesh when exposed to the energy of a high-powered MRI.[41]

VeriChip's Extraordinary Powers?

Even the supposed medical benefits of the chip have not gotten large numbers of people to sign up. Perhaps out of frustration, some distributors have made exaggerated claims, implying that the VeriChip has amazing abilities to make people secure. The most common misperception is that it can prevent kidnappings and locate kidnapping victims. The VeriChip's read range is only a few inches—not the miles that would be needed to locate a missing person. Unless a kidnapping victim passed within eighteen inches or so of a VeriChip reader, the chip would lie dormant without transmitting anything to anyone.*

Some of the misunderstanding may stem from marketers like Applied Digital Solutions' Mexican distributor Solusat, which uses extensive satellite imagery on its website.[42] Or companies like Vinoble, a holding company that ridiculously claims that their "RFID Mobile Location technology," which is "about the size of a grain of rice," offers "corporations, executives, high profile individuals, and any person an added level of safety and security from the threat of terrorist or criminal activity such as kidnapping."[43] Unfortunately, such messages have spread through the culture, leading many to mistakenly believe that the current version of the VeriChip implant can track people from space.

Even though long-distance tracking with the current chip is not possible, there is a more realistic way that people's locations could eventually be tracked through VeriChip-type implants. Applied Digital Solutions markets a doorway portal capable of reading the implants whenever a chipped person walks past. Ultimately, such doorway portals could be installed at strategic locations

---

*While the "grain-of-rice-sized" implant can't track people from space, its parent company has discussed a prototype GPS implant, the size of a pager, that would be surgically implanted under a user's collar bone. This active device would contain a battery and more sophisticated technology that could transmit the wearer's coordinates to a central location. While the company has discussed this prototype for years, no one we know of has ever actually seen one.

around the world, making a log of the comings and goings of chipped people. (Or Applied Digital could piggyback their reader technology into Checkpoint's and Sensormatic's anti-theft portals, already installed at hundreds of thousands of locations.) Such a system would be unlikely to stop kidnappers (who would be smart enough to remove their victims' chips or block their signals before taking victims out in public), but it would make the lives of other chipped individuals an open book.

CHIP REMOVAL

Before this book went to press, we learned that Mexican attorney general Macedo de la Concha had resigned from his post, raising an interesting question. What happens when you no longer need or want the RFID chip?

The glass capsule housing the chips is coated in something called Biobond, a polypropylene substance that encourages the formation of scar tissue around the implant to keep it from migrating. Though chip promoters claim the device is easy to remove, removal would entail far more than a reverse injection process. A surgeon would have to make an incision and cut the chip away from the surrounding tissue to remove it.

While we wouldn't say that sounds exactly easy, a California company called Persephone, Inc., thinks it isn't hard *enough*. They want RFID implants to be almost impossible to remove.

Their proposal? Plant them deeper. Much deeper.

THE INTERNAL ORGAN CHIP YOU CAN NEVER REMOVE

A patent application titled METHOD AND APPARATUS FOR LOCATING AND TRACKING PERSONS lays out Persephone's nightmarish idea: the surgical implantation of their tracking device deep in the body.[44] They've targeted the head, the torso, the deep muscle of limbs, and the lumen of organs like the gastrointestinal tract and the uterus as ideal locations for implantation. There they'd be next to impossible to remove without major surgery.

The patent application explains one of the advantages: "Removal of the implanted device by a runaway juvenile would likely be impossible. Even if possible, such removal would likely place the runaway at significant medical risk, which is counter to the runaway's goal of safe escape and survival from parents or guardians," explains the application.

When not tracking runaways and kidnapping victims, the deep organ implants could have "secondary uses" to track "incarcerated individuals, military personnel, business travelers, [and] mentally impaired individuals. . . ." Someday, they may even include GPS-type implants, so that people could be hunted down "when an activation signal is sent to the implantable device to begin locating and tracking the person."

The device could do more than just track—it could also vibrate, electroshock the implantee, broadcast a message, or serve as a microphone to transmit his or her words to a remote location. Imagine getting an electroshock in your pancreas, or having a microphone inside your head! Here are the relevant excerpts:

Patent Application Publication   Sep. 9, 2004   Sheet 4 of 12   US 2004/0174258 A1

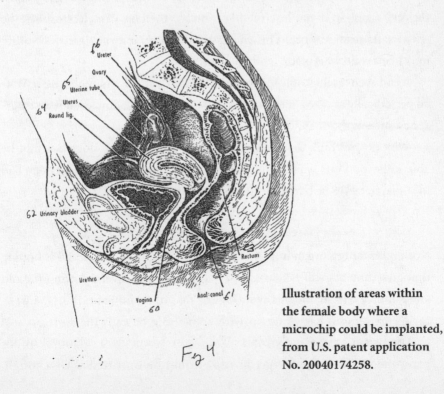

Illustration of areas within the female body where a microchip could be implanted, from U.S. patent application No. 20040174258.

- "Because the device is implanted in the person, it can also provide a shock, vibration, or other warning . . . [that] may be progressive, such that a person is subjected to a shock of increasing magnitude as he leaves a zone of confinement or enters a forbidden zone."

- "An alert can also be broadcast to the person when urgent contact is required. Thus, the device may vibrate or provide other notice when an emergency occurs that requires the person's immediate attention."

- "The device may . . . include a microphone or similar device for monitoring acoustic information, thereby permitting the person to talk to a remote location."

Since the invention can track several persons at the same time, its inventors suggest it would be a great way "to track soldiers on a battlefield, employees within an enterprise campus, or business travelers within a geographic region." Yikes! So if you lose the big contract, the boss can electroshock the entire sales team at once. This makes the Mexican attorney general's office look like a mild workplace.

It can even track people for the heck of it: "The device may also be activated periodically to take a series of fixes on a person's location, even if an emergency, such as a kidnapping, has not occurred."

Why do we think that such a device, if ever mass produced, would be abused beyond our wildest nightmares? Call us cynical, but this is one bad idea that needs to be buried deep—in the dustbin.

THE HUMAN ENSLAVEMENT DEVICE

Not up for a deep organ implant? One future-thinking individual is hoping that governments will take prisoner-tagging to new levels, putting the old ankle bracelet to shame but avoiding the fuss of an implant. Illinois inventor George Vodin has come up with something he calls the METHOD AND APPARATUS FOR REMOTE MONITORING AND CONTROL OF TARGET GROUP.[45] In his plan, members of the "target group" would be outfitted with a tough, tamper-proof armband made of Kevlar bulletproof fabric that would be nearly impossible to remove. But the real kicker is the payload inside—an "injection module" that can be remotely activated to deliver a knockout dose of anesthetics.

Vodin has thought up a lot of applications for his invention. It could be used to restrict people's movements, immobilizing them if they go beyond the bounds set by their captors. An RFID tag in each armband would respond to signals from a "radio gateway" if the person approached an unauthorized area. When triggered, the armband would automatically release its payload into the subject's bloodstream, causing unconsciousness within seconds. Or guards could use a remote control device to instantly knock someone out—just point and shoot, and the person falls helpless to the floor.

The armbands would be particularly effective for controlling large populations. Each would contain a unique ID number that would allow captors to target specific individuals, pinpoint them from a distance (say, through a wireless phone network, which would give them virtually unlimited range), and, with a push of a button, render them unconscious. And if the hypodermic needle isn't enough, the armband can be souped up with "high voltage pulse circuitry, commonly found in tasers or cattle prods" so guards can administer electric shocks, too.

Of course, the "armband" needn't be limited to the arm. It can be redesigned and placed around a person's neck, ankles, or leg. While drug dosages could be calibrated to match different weights and ages, the inventor regrets that it is probably not appropriate for infants, since it may be hard to fit the device to their tiny bodies and "unsafe" to do so. He suggests, instead, imprisoning infants and the elderly in a "separate section with a physical barrier isolating [them] from the armband monitored/secured occupants."

Reading this, it's hard not to think of cattle cars and concentration camps. But if all this talk of government control has got you down, there's another side to Mr. Vodin's invention. He explains that the armbands could be used to monitor patients' vital signs and administer regular doses of prescription drugs. We're still scratching our heads over what sorts of medications would have to be administered through bulletproof armbands, though. ("It's time for your daily dose of apathy, Ms. Albrecht." Bzzzt!)

Or here's one: How about using the anesthetic armbands to keep the skies safe for travel? The inventor suggests that fitted onto airline passengers, the armbands could "automatically activate a drug injection" if a passenger goes "through radio gateway barriers, for example past the passenger area of an airplane towards the cockpit." The captain has turned off the seatbelt sign— but don't take a wrong turn on the way to the bathroom!

We hope this invention will never be implemented. But if some day you find one clamped around your arm, here's a word of advice: Don't try to remove it, bang it forcefully, or block its signal. Stay within range of the readers. And pray the battery doesn't fail. The armband is set to monitor all of these conditions, and will administer "an immobilizing dosage if it detects attempts to remove, isolate or otherwise disable the armband." If the battery does die, its last official act will be to activate the payload and knock you out. While Mr. Vodin limits his discussion to anesthetics like sodium pentathol and ketamine to render you unconscious, it would be just as easy to load the injection modules with something more lethal.

If there were an award for most awful way to use a spychip, we would be hard-pressed to choose between this armband and the deep organ implant/microphone/electroshock device. Slaveholders, third world dictators, sadistic school administrators, terrorist kidnappers, and dungeon prison guards overseeing hell holes everywhere would love to get their hands on one of these—or better yet, on a warehouse full of them.

Who said RFID couldn't be used to control and enslave people?

15

YOUR TAX DOLLARS
AT WORK

The nine most terrifying words in the English language are, "I'm from the government and I'm here to help."

—Ronald Reagan, 40th U.S. president[1]

ID SNIPER RIFLE

Damn! They've got some big mosquitoes around here!" you might exclaim as you slap the spot on your arm where a glass-encapsulated RFID device had just penetrated your flesh. You would never guess you'd just been tagged by government officials wielding the latest high-tech weapon: the Empire North ID Sniper Rifle.

Capable of delivering a GPS-enabled VeriChip at a distance of over a thousand yards, the rifle is promoted by its developer as the ideal tool for "managing and controlling crowds." A government sniper could inject the grain-of-rice-sized microchip implants into persons of interest—say at a protest or demonstration—and then secretly track them by satellite. The best part is that the demonstrators would never know. According to Empire North, the impact of the gun's payload feels "like a mosquito bite, lasting a fraction of a second" and leaves no obvious skin marking. Because the victims would have no inkling of the surreptitious monitoring, it could all be done "without causing damage to the all-important image of the state," the company promises.

Does this sound preposterous to you? It should—especially if you read the last chapter where we explained that not only is the VeriChip bigger than a grain of rice and would hurt like heck going in, it's not directly trackable by satellite. It doesn't have a GPS feature. In fact, it can't even be tracked across a room, since its read range is less than a few feet.

But the Chinese government was a bit more gullible. Rather than question the unbelievable claims of the arms dealer, they bought the concept hook, line, and sinker.

(PHOTO: COURTESY OF JAKOB BOESKOV)

Jakob Boeskov displays his concept of the world's most sinister weapon, the ID Sniper Rifle.

What they didn't know was that Jakob Boeskov, the CEO of Empire North, was no arms dealer. Rather, he was a gutsy Danish artist bearing phony business cards and bogus blueprints to see if governments of the world would buy into a plan to secretly tag and monitor people with the most sinister weapon he could devise.

Surely no one would go for this, Boeskov figured. But when he promoted his weapon at China Police 2002, an international weapons trade show held in Beijing, his worst fears were confirmed. Not only were the Chinese police extremely interested in tracking their citizens with secret microchip implants, but representatives from several other countries expressed an interest, as well.[2]

Imagine that.

Fortunately for Boeskov, no one caught onto his charade or he might have been thrown into a Chinese prison by officials displeased to discover they had been the brunt of the creative caper. But unfortunately for the world's citizens, if Boeskov's weapon did exist, it would likely be embraced and put to use.

GOVERNMENT ABUSE OF RFID

Would the United States government be interested in using a surreptitious, high-powered people-chipping weapon? We hope not. But it's clear that the powerbrokers in Washington have their own plans to promote RFID technology and use it to track and monitor both citizens and visitors to our shores.

Some of these efforts will be announced publicly, like the government-mandated chipping of passports and visitor documents. Others could be virtually invisible to citizens since the bureaucrats won't do the dirty work themselves. They'll leave it to private companies to embed RFID tags into everything, then let retailers associate those numbers with individual consumers. Finally, they'll allow information brokers to record it all in massive databases. That way the government can simply buy the information it wants later, or usurp it outright in the name of national security.

This strategy would give government officials access to the kind of granular, detailed information about us that they want but aren't allowed to collect themselves, thanks to limits placed on them by the Fourth Amendment—the one that protects us from unreasonable search and seizure by the government.

LAW ENFORCEMENT AND SURVEILLANCE

Obviously, accessing RFID tag information would let the government spy on its own citizens to an undreamt-of degree. But what makes us think they'll want to do that?

Their past history.

Back in May 2004, the United States General Accounting Office (GAO) issued a report that documented federal data mining efforts aimed at helping detect bureaucratic fraud, waste, and abuse—a fine and worthy goal. But it also revealed another side: The government uses the same processes to scrutinize us, too. Combing through commercial databases chock full of personal information on consumers, they seek out people that fit certain profiles and target them for close evaluation and scrutiny.[3] In other words, they're using commercial data to go on warrantless fishing expeditions.

Just how widespread are these fishing expeditions? When the GAO surveyed 128 federal departments in late 2003 and early 2004, they found nearly 200 data-mining efforts either planned or underway. Of these, 36 used personal information like credit card transaction data gathered from private sector databases.

Most of us would never volunteer this information to the government unless we were served with a search warrant or someone put a gun to our heads. But the corporations who sell it have no such qualms.

The government is analyzing this information through data mining, a process they describe as using "techniques—such as statistical analysis and modeling—to uncover hidden patterns and subtle relationships in data . . . [to] allow for the prediction of future results."[4]

So the government wants to know not only everything we've done in the past but what we're going to do in the future, too.

What's next, tea leaves and fortune tellers?

This may be just the tip of the iceberg. Several key agencies declined to discuss their data-mining activities with the GAO, including the Central Intelligence Agency (CIA), the National Security Agency, and the Defense Department's Department of the Army.[6] One can only guess at the extent of *their* data-mining activities—or why they chose to keep a tight lid on them. Perhaps the Defense Department has decided it's best to keep such efforts quiet after the public backlash its "Total Information Awareness" program faced a while back.

▶▶ FEDS ACKNOWLEDGE THEY'RE
VIOLATING OUR PRIVACY ◀◀

From the GAO Report on Data Mining, May 2004: "Mining government and private databases containing personal information creates a range of privacy concerns. Through data mining, agencies can quickly and efficiently obtain information on individuals or groups by exploiting large databases containing personal information aggregated from public and private records. Information can be developed about a specific individual or about unknown individuals whose behavior or characteristics fit a specific pattern."[5]

TOTAL INFORMATION AWARENESS

Total Information Awareness, or "TIA" for short, was a Defense Department project designed to capture information about virtually every transaction in every commercial database in the United States. These records—on everything from our phone calls and bank deposits to our store and mail order purchases—were to have been consolidated into centralized government databases where they could be watched around the clock for any unusual activity.[7]

The Total Information Awareness project was de-funded by Congress after a public outcry (some believe it is now operating as a "black bag account," not subject to Congressional scrutiny),[8] but the drive to access increasingly detailed knowledge about us has not gone away. One more terrorist incident and bureaucrats will be clamoring to tear down the remaining shreds of privacy we have left. Of course, unless we're careful, the next time around, they'll get more than our credit card records. They'll get the RFID numbers in our shoes and shirts, too.

WHEN THE DATABASES CONTAIN RFID TAG NUMBERS

If the government gets access to purchase records containing RFID tag numbers, it would be a simple matter to scan the things people are wearing or carrying, look them up in the purchase database, and identify those individuals. This would make it easy to identify political opposition and crack down on civil liberties.

Here's an example. Depending on your politics and interests, imagine you're attending a gun show, a peace rally, a union meeting, a religious service, or a talk by a prominent Muslim cleric. Your right to attend any one of these events is protected by the First Amendment, which guarantees you free assembly with others, so it would be inappropriate for government agents to storm such an event and demand to see ID.

In the spychipped future, however, they could figure out who was there without having to ask. With portable RFID readers in their backpacks, agents could mill around such events, pick up all the RFID tags associated with the people in attendance, cross-reference them in commercial databases, and create a fairly thorough list of who was there.

Not only would that information identify those people, but it could identify their extended web of contacts as well. For example, if Katherine were scanned while wearing her husband's winter scarf, the watch her mother bought her for graduation, and a pen she borrowed from a journalist, the government would know whom to begin questioning about her for more information.

Even if the government could not determine who purchased the objects detected at the event, the unique RFID numbers the objects contain could still pose a threat to their owners' civil liberties. After agents skimmed the numbers from the items associated with a peace rally, for example, they could put *the objects themselves* on a watch list. Even without conveying the identity of the individuals carrying them, the objects could communicate a "peace rally association" (or a gun show association, or a Muslim association, etc.) that government agents could use later.

That way, if a pair of shoes detected at a controversial event later showed up at an airport checkpoint, the wearer could be singled out for further questioning—or even be prevented from boarding the aircraft at all.*

RFID in Financial Instruments

Some have predicted the death of cash, since credit cards and electronic payments are being used more often and for smaller purchases.[9] However, many people (like us!) continue to highly value cash because of the anonymity it affords. This may soon change if RFID takes hold because a method has been developed to track cash and create a history of its ownership trail.

Hitachi has developed a tiny RFID chip called the mu chip, which we mentioned in Chapter Two. A mere 0.4mm square, the mu chip, which contains an integrated antenna, has a read range of only a few centimeters. In spite of its tiny read range, this chip holds the potential to eliminate the anonymity of cash because it is small enough to be embedded in paper currency,[10] and its unique ID number could be captured at any point where cash is transferred.

Imagine if when you took a hundred dollars out of the ATM, each of the twenty-dollar bills you withdrew contained its own unique ID number that could be captured and associated with your account. When you later used one of those bills to make a payment, its number could be captured again by the retailer at the point of sale. If records of these transfers were stored in a master database operated by the federal government (or a private entity that would provide it on demand), it would be possible to literally follow the trail of cash through the economy.[11]

Are any governments using this system? Frankly, we don't know. It was reported a few years back that the European Central Bank was in discussions with Hitachi over chipping EU banknotes to reduce counterfeiting, and it has been rumored that the Japanese government also considered tagging high

---

* Under the CAPPS II passenger screening program, airport screeners don't even have to offer a reason for denying you a boarding pass.

denomination yen notes with the mu chip.[12] However, we don't know if they ever tested or deployed the system.

While there have been Internet rumors that U.S. currency has been tagged with RFID, we have not found any credible evidence to suggest this is the case—though we wouldn't put it past them at some future point.

WHAT ARE THE IMPLICATIONS OF RFID IN CASH?

Of course, once cash is tagged, there will be no more anonymous payment options left—short of bartering (and even that could be problematic if items carry their ownership status in embedded tags). What's more, there will be a tremendous incentive to keep an eye on everyone's financial transactions as a way to reduce or eliminate money laundering, drug dealing, and the black market economy. If money becomes trackable, even giving money to panhandlers and the destitute could land you in trouble. Imagine giving twenty dollars to a homeless man, then having that bill turn up at a drug bust a few days later. Drug enforcement agents could scan the bill and determine that it was issued to you through your bank's ATM machine, and you could find yourself answering some awkward questions.

The read range on RFID tags in cash would be limited to a few inches, given current technology. Nevertheless, astute pickpockets could still pinpoint the easy marks from a crowd by bumping against their pockets with handheld reader devices.

THE UNITED STATES POSTAL SERVICE

Through wind and rain, snow and hail, we deliver spychipped mail. That could be the new United States Postal Service motto if they embed RFID tags in postage stamps as predicted. According to Sun Microsystems, provider of RFID software, the Postal Service is "considering putting RFID capabilities on postage stamps, in order to track and locate mail much more quickly."[13] A unique RFID number on each stamp could one day be used to register every letter to the person mailing it, putting an end to the romance of anonymous love letters and the freedom of correspondence.

A new presidential proposal to grant the FBI authority to track the mail of suspected terrorists could push the timeline up on spychipped stamps. Under the proposal, postal inspectors would be required to supply to the FBI data contained on the outside of correspondence sent to or from subjects of terror investigation.[14] This time-consuming project could be streamlined with RFID-enabled stamps.

In the meantime, the Postal Service is keen to adopt other RFID measures. Their automation consultants have recommended tracking mail containers with RFID tags,[15] and the Postal Service has contracted wireless solutions provider ID Systems to spychip postal vehicles so they can be better monitored and traced.[16] Will postal workers themselves be next?

PASSPORTS

While privacy-conscious consumers had their eye on RFID in consumer products, the U.S. government quietly began an initiative to spychip our passports.

"Sure, the public can boycott Benetton or Gillette, but let's see them try to boycott the U.S. State Department," the Feds must have calculated. "Since we're not elected, they can't even vote us out of office."

Perhaps that was the thinking when the State Department initially forged plans to put unencrypted RFID tags in our passports that would beam stats like our name, place of birth, and nationality to anyone with the right reader device.[17] Americans are already at risk when traveling in countries that oppose U.S. policies, so we're puzzled as to why our government would want to enable kidnappers, thieves, and terrorists to identify us as targets for theft, kidnapping, or worse.

Once again, the decision was made in a bureaucratic context with no opportunity for input from the public. Only *after* the spychipping decision was essentially a "fait accompli" did the State Department open a period for public comment—and it was immediately inundated with thousands of outraged letters of opposition. The torrent of criticism was so relentless that the State Department was forced to reexamine the security issues.[18]

Monitoring Visitors to the U.S.

Foreigners traveling in the United States are being spychipped, as well, thanks to the Department of Homeland Security's United States Visitor and Immigration Status Indicator Technology (U.S.-VISIT) program. Foreign visitors entering the U.S. at certain locations are issued remotely readable spychipped identification devices that contain their name, country of origin, entry and exit dates, and biometric information.[19]

Lest you think this is a good idea, you should know that we're likely to receive a reciprocal welcome from other countries. Do you want to broadcast your nationality when you travel in countries that don't share our government's views on world matters?

Border monitoring could become quite an expensive undertaking—both financially and in terms of advancing the RFID agenda. The government has awarded an unbelievable *ten-billion-dollar* contract to Accenture to develop a "virtual border" for the U.S. Visit initiative.[20] As you may recall, Accenture is the spychip promoter pushing concepts like the "Real World Showroom"—a way to allow strangers to scan you to learn what you're wearing and carrying. This huge infusion of cash will go a long way towards bolstering Accenture's role in domestic affairs. This is worrisome, considering how little attention the company pays either to privacy or propriety.

How will Accenture spend those ten billion dollars? We don't know—and neither does the federal government. Accenture landed the contract (costing taxpayers an estimated forty dollars for every man, woman, and child in the U.S.) without having to specify what they plan to do. But don't worry, there's already talk of iris scans, voice recognition, and digital fingerprinting.[21]

Firearms

The latest RFID security scheme involves implanting a microchip into a gun owner's hand to verify that he or she is authorized to fire the weapon. VeriChip (the implant company) has teamed up with South Carolina firearms manufacturer FN Manufacturing to develop the "smart gun" that would scan the user for an implanted chip before a weapon would fire.

But police officers are not too keen on the idea. Though the "smart gun" is billed as a security measure, the plan has proven unpopular among the very police officers it was designed to protect. A police officer's gun would be useless if the chip in his hand were damaged in a fight, and his partner's gun would be useless to him as well. Worse, there is the potential for savvy criminals to set off an electromagnetic pulse weapon (similar to a high-powered microwave device) to effectively disable the weapons of an entire police force, while still using "old-fashioned" RFID-free weapons to commit crimes themselves.

Relying on an untested technology in a life-or-death situation makes cops understandably nervous. "We have power outages and computer crashes. Would you risk your life knowing all the things that could go wrong?" asked West Palm Beach police training sergeant William Sandman.[22]

Though police officers can still refuse to participate in high-tech gun-disabling schemes, soldiers someday could face a court martial if they say "no." Our men and women in uniform may be the guinea pigs for the military's version of the "smart gun." According to patent application No. 20020149468 (SYSTEM AND METHOD FOR CONTROLLING REMOTE DEVICES), funded by the U.S. government, it is now possible to equip soldiers with weapons that can be

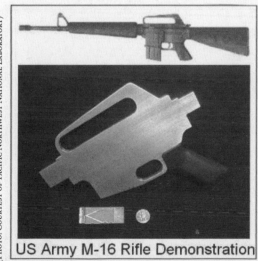

(PHOTO: COURTESY OF PACIFIC NORTHWEST NATIONAL LABORATORY)

US Army M-16 Rifle Demonstration

Government photo shows a gun stock and an RFID tag that might be embedded in it.

remotely disabled courtesy of embedded RFID tags.[23] According to the patent application, "weapons lost on a battlefield can be easily tracked and enabled or disabled automatically or at will." Of course, if the disabling technology falls into enemy hands, our troops could be in a world of trouble. That's a lot of faith to put in a spychip.

It doesn't take much imagination to see how the remote tracking and disabling capability could one day be built into self-defense firearms owned by law-abiding citizens. Imagine if the bad guys, either on a battlefield or breaking into your home, had the ability to disable the good guys' guns. We'd all be sitting ducks. Of course, the government could also use the technology as a convenient way to skirt the Second Amendment and infringe on our right to bear arms.

▶▶ INTERROGATION AND CONTROL FROM THE AIR ◀◀

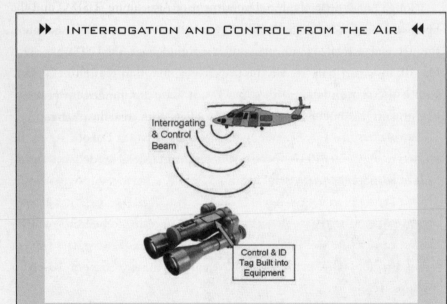

Interrogating & Control Beam

Control & ID Tag Built into Equipment

Dr. Shearer of IBM: Please take note—here's your helicopter! The military envisions RFID tags built into equipment for remote monitoring from the air.

In case you've skipped ahead, you need to go back to Chapter Thirteen and read the slur about helicopter-monitoring to fully appreciate the significance of this graphic. (PHOTO: COURTESY OF PACIFIC NORTHWEST NATIONAL LABORATORY)

The Government Promotes RFID

Far from protecting the public from the RFID threat, our government is actively promoting the technology. Some agencies have been on board from the start, such as the Department of Defense and the United States Postal Service, both "sponsoring members" of the Auto-ID Center from its earliest days. And as far back as 2002, the Auto-ID Center was already meeting with Office of Homeland Security Director Tom Ridge.[24]

The Defense Department, one of the largest purchasers of goods in the world, made a major tactical move to advance the spychip agenda in September 2003 by requiring its suppliers to affix RFID tags onto shipments headed to DOD warehouses.[25] As we've said, it was the government equivalent of the Wal-Mart mandate.

Not to be outdone, the U.S. Department of Agriculture (USDA) and the Food and Drug Administration (FDA) are both encouraging RFID adoption through their recently issued "track and trace" guidelines. The USDA has called for all food animals to be tracked from "birth to slaughter"[26] (see www.NoNAIS.org for more), while the FDA, as we've explained, wants prescription drugs tracked from the manufacturing plants through to the pharmacies.[27]

Lawmakers like U.S. Senator Byron Dorgan of North Dakota are in the pockets of big money RFID interests. Dorgan spent federal tax dollars to bring RFID tag manufacturer Alien Technology into a partnership with North Dakota State University to support "micro-technology research." It was little surprise when Alien later returned the favor by agreeing to build the world's highest capacity RFID chip manufacturing plant in Dorgan's state, bringing lots of dirty jobs into North Dakota.[28] Welcome to the new economy based on spychips.

By the end of 2004 and the beginning of 2005, major RFID initiatives were publicized by a number of government agencies, including the Social Security Administration,[29] NASA,[30] the Postal Service,[31] and the Department of Homeland Security,[32] among others. The volume and fury of these announcements reached such a fevered pitch that we knew something big was happening behind the scenes.

It all became clear when we found a General Services Administration (GSA) bulletin titled "B-7 Radio Frequency Identification," blatantly directing heads of federal agencies "to consider action that can be taken to advance the [RFID] industry by demonstrating the long-term intent of the agency to adopt RFID technological solutions."[33] Hold on! Advance the industry?! The directive was signed on December 4, 2004.

Apparently, since the spychippers knew they couldn't get consumers to sign on, they switched tactics to getting government agencies on board instead. Taxpayers have little say in the day-to-day purchasing decisions of federal agencies, so it was the perfect place to land high-profile and lucrative contracts—right over the heads of the public. After all, how is a citizen going to protest the Social Security Administration's use of RFID to keep track of files? Not only could government agencies spend freely to support the RFID industry, but also their deployments would make it clear the technology had government-backed legitimacy. Quite a coup.

U.S. Senators Vow to "Protect" Spychips

As we know from the confidential documents we uncovered, the RFID industry has long been planning to use "top tier" government officials to advance its agenda. Apparently, those efforts are now paying off. Not only is the GSA openly supporting RFID, but U.S. senators are getting on board, too. Rather than looking out for the interests of their constituents, our elected representatives are working overtime to protect and promote the interests of the RFID industry.

One pro-RFID government missive to leave us reeling came from something called the "Senate Republican High Tech Task Force," which unveiled a set of policy programs in the spring of 2005. Unbelievably, one of their policy planks was a vow to "protect" RFID. These senators announced they would:

> protect exciting new technologies from premature regulation or legislation in search of a problem. RFID holds tremendous promise for our economy, including military logistics and commercial inventory efficiencies, and should not be saddled prematurely with regulation.[34]

We were disturbed when we took a closer look at this "task force" of elected officials. Its website describes it as "a conduit for the technology industry."
But, wait a minute. We didn't elect these senators to represent the technology
industry; we elected them to represent *us*. When did politicians become lack-
eys for industry instead of "conduits" for the people? CASPIAN doesn't generally advocate legislative controls over RFID (we want labeling legislation only
as we describe in Chapter Seventeen), but we certainly don't think it is appropriate for our elected representatives to gush about the technology, calling it
"exciting," either. We're betting the lawmakers' exuberance will subside once
their constituents read this book and learn that the "exciting new technologies"
their politicians are pushing involve Orwellian-style privacy invasions.

16

THE NIGHTMARE SCENARIO

Power kills; absolute power kills absolutely.

—R. J. Rummel[1]

Sed quis custodiet ipsos custodes?
[Who will watch the watchers?]

—Juvenal, *Sixth Satire,*
first century A.D.

THE HUNTER OR THE PREY?

There are two ways to look at RFID's extraordinary human tracking abilities. Either you embrace the idea of being able to identify and track individuals everywhere they go or you recoil in horror at the thought. Your reaction depends on where you see yourself in the tracking equation: Are you the hunter or the prey?

Often, when we give an RFID talk to a group of corporate executives, a strange transformation takes place. The same pro-RFID folks who drum their fingers on the table and shoot us icy looks before the talk begins will leap terrified out of their seats when we're through, saying, "How do we stop this? We can't let them do this to us!" It's funny how easily "we, the watchers" can turn into "we, the watched." In other words, RFID is fine if we're the ones wielding it, but not if somebody else is at the helm.

Nowhere is the predator mentality more clearly articulated than in a 2001 promotional video featuring former MIT Auto-ID Center executive director Kevin Ashton (the visionary from Procter & Gamble who started this whole thing). As you read the following story by Ashton, see if you identify with the predator or the prey:

> I was in Africa last year, and we were in a Jeep tourist thing. [We] came across some zebra, and our very amusing African tracker turned to me, the MIT guy, and said, "Okay, I bet you don't know why zebra have stripes." And, of course, I knew the answer to this; it's obviously camouflage, isn't it? When the guy stopped laughing (which took some time), he asked me, "Okay. Zebra. Black and white stripes. Africa. Do you see anything black and white? No."
>
> He told me zebra have stripes for a very interesting reason. And the reason is that if a predator decides to attack a herd of zebra, what it wants to do is singulate one, hunt it down and exhaust it. The reason a zebra has stripes is so that when the zebra all run away, it's absolutely impossible to keep your eye on any one individual. So the stripes on a zebra are actually a defense against identification.
>
> That caught my imagination because of what I do for work, where identification is the central problem. In nature, identification is a matter of life and death. If you can't identify things, you can't count them, you can't work out whether or not you can eat them, you can't work out whether or not they are friends or foe. [2]

In his usual insightful way, Ashton has neatly summed up the issue. Without a failsafe method of identification, a predator can't single out an individual, hunt it down, exhaust it—and ultimately eat it. If you're a lion, you should embrace RFID. If you're a zebra, you should fight like hell to keep it away from yourself and your children.

Today, our public discourse is almost entirely dominated by the information predators—the lions who want to keep a close watch on everyone else. Usually this is framed as a battle where the "good guys" need better technology to watch the "bad guys." The chorus can be deafening.

"Stop the terrorists!"

"Photograph the shoplifters!"

"ID the foreigners!"

"Tag the truant kids!"

We are inundated with so many messages from the watchers that their worldview begins to seep into our consciousness, drowning out our awareness of a crucial fact: Once we empower the lions, it won't take much to become their prey ourselves.

Imagine what it would mean for society if RFID actually lived up to its promise and made it possible for authorities to single us out as individuals and have 100 percent, round-the-clock accuracy about who we are, where we go, whom we associate with, and what we do with our time. Imagine that ubiquitous RFID, coupled with omniscient databases and an all-seeing video surveillance grid, made law enforcement so knowledgeable that no act could ever go unobserved and no crime would go unpunished.

Maybe this sounds good to you. It would almost certainly cut down on terrorism, theft, kidnapping, and even petty street crime—all positive outcomes. If you identify with the lions, it's a good thing for law enforcement officers and Homeland Security officials and border guards and school principals and everyone else in authority to have absolute omniscience.

That is, of course, unless they've gone bad.

THE YELLOW STAR

In the dark days of Germany's Third Reich, Jewish people had become the hunted. The only way they could escape deportation to Nazi death camps was by fleeing the country, hiding, or blending in with the rest of the German herd, using illegal means like forged identity papers and elaborate cover stories. Tragically, many failed. But some Jews used their zebra stripes to avoid being singled out from the crowd and targeted for destruction.

In his book, *The Last Jews in Berlin*, author Leonard Gross recounts the miraculous survival stories of a handful of Jews who rode out World War II from within the borders of Germany itself. Their strategies included going underground or posing as non-Jews to escape detection. But regardless of how they survived, these rebellious, brave souls had all taken a common first step: They discarded the yellow star that identified them as Jews and marked them for death.

Interestingly, after years spent interviewing Holocaust survivors for his book, Gross arrived at the same "herd" analogy that Kevin Ashton would later use to justify RFID. The crucial difference is that where Ashton identified with the hunters, Gross identified with the hunted. Here he describes how Nazis singled out their prey from the herd:

> On September 1, 1941, the Nazis had ordered all Jews older than six to wear a Star of David over their hearts. . . . It was a yellow star outlined in black and embroidered with the word *Jude*. . . . They were forbidden to leave their districts without permission or to be outdoors after evening curfew hours— policies whose underlying purpose became clear once the deportations began. Not only had the Jewish cattle been branded for easy identification, they had been penned into stockades where their captors could cut them out of the herd for the trip to the slaughterhouse.[3]

The "final solution" had been a long time coming. Even before the death camps and extermination began, the Nazis had turned Jews into nonentities in their own communities, doing everything they could to prevent Jews from enjoying the conveniences and pleasures of life. As Gross explains, Jews were barred from using public parks and public streets where official buildings stood. They were banned from using public telephones or restrooms. Even public transit was off limits, except in certain, limited circumstances. Even harder to abide, they faced draconian restrictions at stores. Jews were prohibited from buying basic staples, including eggs, milk, cheese, white bread, smoked meats, fish, tobacco, and spirits.

There's little doubt in our minds that were the Holocaust to happen today, the Nazi predators would have done more than issue yellow stars to mark their victims. They would almost certainly have tagged every Jew with a mandatory RFID implant, preferably deep in the body where it would be next to impossible to remove.

In a world filled with RFID readers, the Nazis could have been far more efficient in depriving Jews of access to basic necessities and the stuff of daily life. RFID numbers encoded in their chips could mark Jews as social and technological pariahs, causing any doorway, elevator, or appliance equipped with RFID-based authentication to shut down when a Jew attempted to use it. In a cashless society where an ID swipe is required for nearly every activity, pay phones could be programmed to withhold dial tones, subway gates could remain firmly closed, and store equipment could refuse to ring up "Aryan-only" food like eggs and milk for the "wrong" kind of person. A few keystrokes could cut off an entire community from the herd, in a horrific mutation of the marketers' "digital redlining" technique. But rather than charging higher prices and offering poor service to the "undesirables," such techniques could prevent them from receiving any services whatsoever.

When the low-tech world goes bad as it did in Nazi Germany, it's a nightmare, but when the RFID world goes bad, the nightmare could permeate every aspect of its victims' lives, making camouflage and escape all but impossible. RFID could fulfill dictators' wildest evil dreams, providing near total omniscience and control over every aspect of society.

When RFID goes bad, it will be unlike anything we've ever seen before.

COULD IT HAPPEN HERE?

Perhaps you think talk of government abuse is a red herring, since, after all, we're not living in Nazi Germany. That's probably what Germany's neighbors thought, too, until fascist thugs grabbed the reins of power from their legitimate governments and began committing the same atrocities they had done at home. Had countries like Poland or the Netherlands built up an RFID surveillance infrastructure, regardless of how benign their intentions and how many legal controls they had put in place to constrain its use, the Nazis would have seized it, removed the safeguards, and quickly applied it to their own abhorrent ends.

It would be so easy for an oppressor to dominate a population accustomed to being watched and controlled by RFID tags and readers in their homes, schools, stores, and workplaces. A people docile enough to allow their medicine cabinets, refrigerators, cash registers, retail shelves, food, guns, passports, mail, work uniforms, car tires, roads, taxis, and subways to be tagged and monitored by authorities would be easy pickings for a tyrant. If a people can't even fight back against snoopy marketers and their own elected representatives, how would they fare against an armed and aggressive enemy?

That's the problem with power, and why total government omniscience is a bad idea. No matter how much you trust your government, giving it unchecked ability to observe and control your life is like putting a noose around your neck and hoping the guy on the other end never pulls the rope. You might think you're handing that rope to Mother Teresa only to find yourself one day staring into the eyes of Lyndie England.[4]

Or Adolf Hitler.

GOVERNMENT: THE DEADLIEST FORCE ON THE PLANET

Still think it couldn't happen to us? The rise of a bloodthirsty government like the Third Reich is anything but an isolated event. Government violence is a timeworn reality that stretches back throughout history. University of Hawaii professor R.J. Rummel has devoted his career to researching this phenomenon, which he calls "democide," the killing of people by their own governments. What he has found is breathtaking—and very scary.

Hundreds of millions of people have been slaughtered in cold blood by the very authorities that were supposed to be in charge of protecting them. In fact, in the twentieth century, people's own governments were four times more deadly than all the century's wars combined. Rummel cites examples of democide from around the globe—from China, the Soviet Union, Germany, Portugal, Mexico, Japan, Vietnam, Indonesia, Poland, Pakistan, Turkey, Cambodia, North Korea—the list is mentally and emotionally exhausting.

20TH CENTURY DEMOCIDE

REGIMES	YEARS	(000) TOTAL KILLED
MEGAMURDERERS	1900–87	151,491
DEKA-MEGA	1900–87	128,168
U.S.S.R.	1917–87	61,911
China (PRC)	1949–87	35,236
Germany	1933–45	20,946
China (KMT)	1928–49	10,075
LESSER	1900–87	19,178
Japan	1936–45	5,964
China (Mao Soviets)	1923–49	3,465
Cambodia	1975–79	2,035
Turkey	1909–18	1,883
Vietnam	1945–87	1,670
Poland	1945–48	1,585
Pakistan	1958–87	1,503
Yugoslavia (Tito)	1944–87	1,072
SUSPECTED	1900–87	4,145
North Korea	1948–87	1,663
Mexico	1900–20	1,417
Russia	1900–17	1,066
CENTI-KILOMURDERERS	1900–87	14,918
TOP 5	1900–87	4,074
China (Warlords)	1917–49	910
Turkey (Ataturk)	1919–23	878
United Kingdom	1900–87	816
Portugal (Dictatorship)	1926–82	741
Indonesia	1965–87	729
LESSER MURDERERS	1900–87	2,792
WORLD TOTAL	1900–87	169,202

This table compiled by University of Hawaii professor R.J. Rummel provides a grim accounting of government murder in the twentieth century.[5]

Though these countries had widely different cultures, languages, and geography, Rummel found that they all shared one common feature: *excessive government power.*

In total, during the first eighty-eight years of [the 20th] century, almost 170 million men, women, and children have been shot, beaten, tortured, knifed, burned, starved, frozen, crushed, or worked to death; buried alive, drowned, hung, bombed, or killed in any other of the myriad ways governments have inflicted death on unarmed, helpless citizens and foreigners. The dead could conceivably be nearly 360 million people. It is as though our species has been devastated by a modern Black Plague. And indeed it has, but a plague of Power, not germs.[6]

What are we to make of these statistics? Rummel's conclusion goes to the heart of the problem. He writes that "the way to end war and virtually eliminate democide appears to be through *restricting and checking power*" (emphasis added).

SURVEILLANCE IS POWER

Governments like to assure their citizens that surveillance will make them safer, but surveillance is more likely to ensure the security of the regime in power than to protect the citizens. Once surveillance tools are in place, governments are tempted to use them to identify and hassle people who oppose their rule, whether they are members of opposing political parties (think Watergate) or citizens acting for peaceful change (think Martin Luther King, or, more recently, twenty-one-year-old Sara Bardwell, a member of the group "Food not Bombs" that cooks for the homeless, who was intimidated by the FBI for protesting the Iraq War[7]). Surveillance by the state has a chilling effect on people's willingness to work for social change and root out abuse. In a surveillance state, people keep their heads low and conform. And, of course, that's just how the government likes it.

Remember the all-seeing surveillance grid that sounded so good a few pages back? Here's what we wrote:

Imagine what it would mean for society if RFID actually lived up to its promise and made it possible for authorities to single us out as individuals and have 100 percent, round-the-clock accuracy about who we are, where we go, who we associate with, and what we do with our time. Imagine that ubiquitous RFID, coupled with omniscient databases and an all-seeing video surveillance grid, made law enforcement so knowledgeable that no act could ever go unobserved and no crime would go unpunished.

▶▶ DOES SURVEILLANCE KEEP PEOPLE SAFE? ◀◀

One of the most surveilled people in all of history were the Soviets under communist rule. During Stalin's decades-long reign of terror and the KGB era that followed, government agents could intercept and read mail, listen in on phone calls, and plant informants to probe their neighbors' political views and assess their loyalty to the state. The surveillance was near complete, but did the watchful eye of the state keep the Soviet people safe? Hardly. It seems no coincidence that history's most watchful regime was also one of its most deadly. Between 1917 and 1987, the Soviet government killed over sixty million of its own citizens—more than any other government in the twentieth century.

Considering Rummel's democide statistics, wouldn't you rather take the better odds of risking random crimes than face the possibility of total control by a bloodthirsty government? No criminal we know of has managed to murder an average of three thousand people *per day, every day,* for seventy years, as the Soviet government did.

It would take an army to kill seven million people in a single winter, as Stalin did with a state-induced famine in the Ukraine in 1932–33. Stalin first confiscated all of the Ukrainian's seeds so they couldn't plant crops or store food for the winter. Then he sent troops to search barns and cellars for hidden grain or hoarded food. Finally, when winter came, he ordered the military to close the borders and prevent food from reaching the people. At one point, Ukrainian villagers, Stalin's subjects, were dying at the rate of twenty-five thousand per day—that's more than one thousand people per hour, seventeen people per minute.[8]

These statistics alone should give us pause before we implement an RFID infrastructure that could allow the government to monitor and control everything, including our food.

Challenging Our Basic Assumptions about Reality

We are standing on the brink of a new era. When RFID proponents refer to their new technology as "disruptive," they are making a breathtaking understatement. RFID has the power to change our lives in fundamental ways, undermining our basic assumptions about the world around us.

Up until now, maintaining a line between public and private knowledge of our physical possessions has been as intuitive and simple as breathing. If someone else can't see something, can't hear it, smell it, touch it, or taste it, they won't know it's there. We have always maintained our privacy through simple physical acts like putting something into a box or a bag where we know it can't be seen. Every time you wrap a present, tuck a letter into a drawer, close a door, put money under a mattress, or slip something into your pocket, you are relying on this basic assumption. It underpins our notions of safety and physical privacy.

By creating a form of x-ray vision that can see through pockets, walls, and wrapping paper, however, the spychippers hope to change all of that. Their technology opens the door to a "transparent society" where everything we do can be monitored, scrutinized, and observed by others. In the future, even retreating to your home and locking your door might not shield you from the prying eyes of the world outside. The end result could be just as damaging to the social world as mishandled nuclear energy to the physical world. And like nuclear energy, the fallout could take years to fully recognize.

The RFID industry is not taking the threat seriously. When someone is playing with fire, you want them to acknowledge that fire is dangerous, since only then will they be likely to take commonsense precautions. If scientists set up a nuclear test lab in a suburb, for example, we'd sure feel better if the people in charge acknowledged the dangers of radioactive material and encased the building in lead. But frighteningly, the developers and promoters of RFID are behaving more like Alfred E. Neuman—"What me, worry?"—caricatures than responsible stewards of our future.

Rather than think hard about the world-changing power they are about to unleash on society, the RFID peddlers say, "We don't need to take precautions because there is absolutely no danger here." Then, they pour money into PR campaigns to "pacify" the rest of us and lobby government officials to shield and promote their technology.

There will always be those who believe the potential societal benefits of surveillance schemes outweigh the risks of abuse. However, though there is ample evidence that the supposed security "benefits" of mass surveillance are quite doubtful,[9] the risks of unchecked government control are very real and not to be discounted. As the police and other agents of the state increasingly tap the power of the retail sector's growing arsenal of sophisticated surveillance technologies, we may soon find ourselves in the totalitarian nightmare described by George Orwell in *1984*.

The Slow Progression

Now that we have the ability to do it, the pressure to require some form of permanent, foolproof ID for every person on the planet is bound to increase steadily. We're already seeing the beginning of this mandate with the Real ID Act recently passed by Congress. The starting point will be spychipped driver's licenses, building access cards, and student name tags, and the end point will be microchips embedded in our flesh.

Unless we act now, it's just a matter of time before society finds a compelling reason to permanently identify and track "captive" populations with implantable microchips. First, we'll implant society's outcasts—like prisoners and the homeless—justifying it as a security measure. When such chipping becomes commonplace and hence "acceptable" in those populations, society may expand those efforts to semi-captive populations like the elderly, school kids, and the military. Next will come government employees and those working for major corporations. After all, the argument will go, no one's forcing you to do it—although if you don't go along, you can kiss your paycheck goodbye. Finally, when most everyone else has been signed up, they'll start coming for the rest of us. Nicely at first, then in earnest.

Not chipped? Is there a problem? Don't you realize you are putting us all at risk?

Kevin Ashton: "We Will Have to Die"

In the same video where Kevin Ashton so clearly defines the crux of the RFID argument with his zebra metaphor, he theorizes what will have to happen for RFID to be accepted by society. Without skipping a beat, the always cool-as-a-cucumber Ashton tells gathered pro-RFID executives, "We will have to die."

After awkward laughter from his audience, Ashton clarifies the seemingly outrageous statement. Our generation, he explains, will never fully embrace a world where everything can be tagged and tracked. It's just too new. But the next generation will.

Adolf Hitler understood this dynamic well, saying:

> When an opponent declares, "I will not come over to your side, and you will not get me on your side," I calmly say, "Your child belongs to me already. A people lives forever. What are you? You will pass on. Your descendants, however, now stand in the new camps. In a short time they will know nothing else but this new community."[10]

It is up to us to protect our children's generation from accepting unbridled government power and surveillance. If we embrace spychips today, our children and grandchildren will grow up trained to report their every move, activity, and purchase—even the contents of their backpacks and purses—to marketers and government officials. If we do not nip this trend in the bud now, while there is still time, our grandchildren may not know what privacy and anonymity are. Every time they step through a doorway, attend a class, enter a library, or even walk in the park, a computer somewhere will be watching.

Fight back!

[T]he [Nazi] revolution arrived not with a rush but covertly and, at times, even comically. There were no battles to fight, no bastilles to storm. Men and women fell into the arms of the new Reich like ripe fruit from a tree.... [T]he Nazi revolution was orderly and disciplined. But the reason lies not so much within the Nazis themselves as in the lack of an effective opposition. . . . [M]illions watched passively, not deeply committed to resistance.

—George Mosse, *Nazi Culture*[11]

The RFID revolution planned by global corporations and governments will be nearly imperceptible at first, as the technology slowly permeates warehouses, then spreads to store shelves, our homes, and then perhaps ultimately into our flesh. Because of its silent and insidious nature, the spychipping infrastructure could be in place before we even have a chance to weigh in on its development.

Can we afford to cede our anonymity in the illusory hope that the RFID network will never fall into the wrong hands? It is up to each of us to ensure that comprehensive, all-knowing surveillance systems are returned to the scrap heap of history's bad ideas before it is too late to stop them.

PULL THE PLUG!

Nobody made a greater mistake than he who did nothing because he could only do a little.

—Edmund Burke[1]

Find out just what any people will quietly submit to and you have found out the exact measure of injustice and wrong which will be imposed upon them, and these will continue till they are resisted with either words or blows, or both. The limits of tyrants are prescribed by the endurance of those whom they oppress.

—Frederick Douglass[2]

A MESSAGE OF VICTORY

We don't have to feel hopeless, outnumbered, or discouraged in the face of the RFID threat. The good news is that businesses depend on our shopping dollars, and this gives us powerful leverage. If consumers don't want spy-chips—*and act on that preference in the market*—companies will stop using RFID, plain and simple.

This is not a pipe dream. When they hear of RFID, two thirds of people—an extraordinary majority—immediately understand what the technology will mean for their privacy and oppose it.[3, 4, 5] Despite media messages to the contrary, ordinary people care deeply about their privacy and are ready to take a stand against the erosion of their rights. Americans will not surrender their cherished freedoms without a fight—and they certainly won't trade them for five cents off a bag of rice or sacrifice them to save a few minutes in a checkout line. We know this because we have received literally tens of thousands of e-mails from people concerned about their privacy who tell us they are ready to take a stand. "No more surveillance. Enough is enough," they say. We can get this message out through the power of our spending choices and by banding together and speaking up.

We can use the power of the market to put an end to spychips on consumer goods. The first step is to identify the companies that are using RFID irresponsibly and refuse to shop in their stores or buy their products. We have listed the most notorious of these spychipping companies in the box below, and we'll keep an updated list at Spychips.com.

The next step is to get as many other people as possible to boycott the products. Talk to neighbors, family, coworkers, and friends. You can also use protests, websites, flyers, bumper stickers, posters, town meetings, and more to help spread the word to others.

Businesses can choose to respond to our demands for spychip-free products or not, but the market will punish those who fail to pay attention to consumer concerns.

Of course, the flipside of punishing corporations that behave badly is lavishing rewards on those that respect our privacy and treat us with dignity. We must pledge our business to companies that take a public stand against item-level RFID tagging and promise their products will be spychip-free. This is the beauty of the free market. When it works correctly, both parties are happy with the relationship and everyone benefits.

THE RFID RIGHT TO KNOW ACT

Boycotts are fine and well, but what about the government? Shouldn't they pass laws to protect us from RFID privacy invasion?

In today's legislation-heavy climate, people immediately look to the government to protect them from marketplace threats. But there are two problems

▶▶ THE WORST OF THE SPYCHIPPERS ◀◀

The following companies have past, present, or future plans to use—or abuse—RFID on consumer products.

Gillette: Co-founded the Auto-ID Center. Hid spychips in Mach3 razors and installed a "smart shelf" to snap secret photos of customers at a Tesco store in England. Installed a similar "smart shelf" in Brockton, Massachusetts. (See BoycottGillette.com for details.) Gillette VP Dick Cantwell, the former chairman of the Auto-ID Center, continues to aggressively promote item-level tagging of consumer goods. Gillette is slated to merge with Procter & Gamble (below) to create the world's largest consumer goods company. Products: Gillette razor blades, Oral-B, Duracell, Braun.

Procter & Gamble: Co-founded the Auto-ID Center with Gillette. Hid spychips in Max Factor Lipfinity lipstick being sold at a Wal-Mart store in Broken Arrow, Oklahoma, then secretly videotaped women interacting with them. Denied the trial until proof surfaced. Designed invasive "Home of the Future" and "Store of the Future" prototypes. Continues to vocally promote item-level RFID tags. *Products:* Max Factor, Crest, Cascade, Tide, Clairol, and more.

Wal-Mart: Issued the now infamous RFID supply chain "mandate" to force its top one hundred suppliers to invest in RFID. Within a year, the mandate had prompted hundreds of millions of dollars of investment in RFID infrastructure, launching the nascent industry. Wal-Mart worked with P & G to videotape women interacting with spychipped products. Installed an RFID "smart shelf" in Brockton, Massachusetts, then denied it had done so. Currently, Wal-Mart is selling item-level tagged Hewlett Packard products in seven Dallas-Ft. Worth, Texas, stores, in violation of a call for a moratorium on item-level tagging issued by the world's leading privacy and civil liberties experts.

Tesco: Currently putting item-level RFID tags on consumer goods, including DVDs, for sale at stores in England. This use of RFID is in violation of a call for a moratorium. The tags are not disabled at the point of sale. Participated in "smart shelf" trial with Gillette to photograph customers picking up razor blades. (See BoycottTesco.com for details.)

Metro/Kraft/Nestlé/Johnson & Johnson/Henkel/P&G/Gillette/IBM: All involved with the notorious "Future Store" in Rheinberg, Germany, where customers were issued store shopper cards with RFID tags hidden inside, and tag deactivators didn't work.

IBM: Holds several nightmarish patents detailing ways to use RFID to spy on people. Has referred to customers as "guinea pigs."

Checkpoint Systems and Sensormatic: These anti-theft companies' item-level RFID "source tagging" programs violate our call for a moratorium and violate EPCglobal's own "privacy guidelines." Their "Liberty" portals could usher in an undetected RFID surveillance infrastructure.

Other Companies to Watch: Abercrombie & Fitch, Marks & Spencer, Levi-Strauss, Ameripride and Cintas uniform companies.

with asking lawmakers to regulate RFID. First, there is the lack of political will to do it, as we pointed out in Chapter Fifteen. Corporate lobbyists have already begun whispering into politicians' ears how much more profitable it will be for them to "protect" RFID than to regulate it. But even if we *could* get lawmakers to control RFID through legislation, it would be a bad idea. The reason goes straight to the heart of the new consumer movement: If we're going to fix this mess, *we'd better start wielding power ourselves.*

Relying on the government to preserve our freedom and privacy is like asking a troop of foxes to preserve our hens. It's simply not in their nature to do it. While *the people* will always seek freedom and privacy, government is the natural enemy of these aims—a fact the founding fathers of our nation understood well when they crafted ways to limit the government's power.

Begging the government, hat in hand, to solve our privacy problems for us is not only humiliating and ineffectual, but it turns us into a bunch of weak-willed sycophants, too cowed and domesticated to do anything for ourselves. We have to stop petitioning the corridors of power on bended knee, asking favors they are unlikely to grant, and instead *ourselves* become a powerful force to be reckoned with. This is how people throughout history have over-thrown tyranny and regained their liberty—not by asking the usurpers nicely for their rights but by standing up and claiming them.

We believe the only appropriate role for RFID legislation is to require that companies tell us whether or not products contain RFID tags so we can make our own informed decisions about whether or not to buy them. Since spy-chips can be so easily hidden, it's possible that even the savviest RFID oppo-nent could accidentally buy a product or clothing item containing one. To prevent this, we have developed model legislation that would require items containing RFID to be clearly labeled. This legislation, the *RFID Right to Know Act*, is available at the Spychips website.

The RFID Resistance Is Growing Strong

We consumers have had an enormous impact on limiting RFID abuses, and we can win the next phase, too. With only a shoestring budget and a staff of unpaid volunteers, we've successfully faced down companies like Benetton, Gillette, Procter & Gamble, Wal-Mart, Metro, and Tesco—the Goliaths of the corporate world—and gotten them to back away from RFID deployments.

▶▶ WE'VE BEEN HEARD! ◀◀

"From Wal-Mart stores in California to the Metro supermarket in Germany, CASPIAN and others have forced retailers to feel their pain or feel their wrath. Expect further concessions as privacy advocates flex their muscle."

—*Business Week*, March 2004[6]

Recall that Benetton canceled its clothing tagging plans just weeks after our boycott was launched, Gillette's photo-snapping "smart shelves" disappeared from Wal-Mart and Tesco stores overnight, and Procter & Gamble has kept a low profile since their secret lipstick tagging webcam scheme was discovered.

And CASPIAN is not alone. We've been joined by dozens of the world's most prominent privacy and civil liberties organizations in condemning item-level tagging and calling for a more open debate on RFID.

Citizens have begun to fight back against the government's plans to use spychips, too. In the spring of 2005, Bill Scannell spearheaded a campaign at his RFIDKills.com website to oppose the RFID tags being planned for our passports. In just days, the website generated over two thousand letters to the State Department as people from all over the nation wrote to voice their opposition. In response, the government announced that it was rethinking security issues associated with the RFID-chipped passports it had planned.[7]

In San Francisco and Berkeley, Peter Warfield of the Library Users Association and Lee Tien of the Electronic Frontier Foundation (EFF) have turned up the heat on public libraries wanting to spend hundreds of thousands of dollars to spychip books. And, of course, we can't forget the courageous families who just said "no" when the Brittan School in Sutter, California, wanted to hang spychipped tracking cards around students' necks. Together with privacy organization EPIC and the ACLU, they forced a hasty end to the program.

Opposition to RFID spans the globe. "The Position Statement on the Use of RFID in Consumer Products" has been translated into several languages, including Spanish, German, and Japanese. Organizations like Liberty, the National Consumer Council, and Privacy International have all weighed in on the RFID debate, educating consumers and decision-makers in the UK. And our UK sister organization, NoTags, has protested Tesco's spychipping and spread the word through the British media.

Privacy advocates in Spain, France, and Australia have waged educational campaigns, and, of course, a coalition of German consumer groups led by

FoeBuD braved a nasty snowstorm to protest at the Metro Extra Future Store. Their overwhelming opposition forced Metro to recall ten thousand spy-chipped loyalty cards.

Finally, we're encouraged by the fact that the people involved in the RFID industry are themselves consumers and parents, too. While we may be tempted to think of them as slavering control freaks (and some of them clearly are!), the truth is that most people implementing RFID systems today are focused on earning a living. Most have never really thought about the societal implications of what they do or considered that the same technology that allows them to track a warehouse full of paper towels could someday be used to enslave future generations. Once they learn the truth about RFID, we believe that many people within the industry itself will begin to demand better accountability and societal safeguards on the technology.

You Can Fight RFID, Too!

One of the most common questions we hear from new CASPIAN members is "What can *I* do to make a difference?" The answer: Lots!

You can start small with a few easy actions, or dive into some serious activism with both feet. We've compiled a list that offers a range of activities so you can pick and choose what you feel is right for you. You may want to start small and get your big toe wet a few times before jumping into the deep end of the pool with us.

Small Steps

Following are some first steps every concerned consumer can take to protect his or her privacy and take a part in the fight against RFID:

Avoid buying products from companies that promote RFID. By withholding your shopping dollars from companies that support spychipping, you'll be exerting market pressures on them to behave responsibly and sending a powerful message to others, as well. (Remember, the Spychips website keeps an updated list of companies known to be involved with RFID.)

Pay cash for purchases. Paying for your purchases with anonymous cash (and without submitting a form of identification, like a loyalty card) is one of the best ways to protect your privacy. Using cash will also help send a message that we need and want to use cash. If we don't use it, we could lose it.

Do not shop at stores that require frequent shopper or "loyalty" cards. As we pointed out in Chapters Five and Fifteen, that's akin to opening your door to marketers and government agents to let them know exactly what you've purchased over time, painting an intimate portrait of your life. Instead, seek out privacy friendly stores that do not require a card.

Pay cash at the toll booth instead of using an automatic toll transponder, like a "FasTrack" or "EZ-Pass" automatic toll payment device. As we point out in Chapter Eleven, those toll tags can be tracked miles from the toll booth. If you must use a toll tag for roads that offer no other option or because of the hefty premium charged for cash payment, consider keeping your toll tag in an RF shielded pouch when you're not paying a toll. Someone has suggested attaching the toll tag to the windshield with Velcro so it can be removed at will.

Give up something you love. Giving up just one product manufactured by an RFID-promoting company can have an impact and give you the satisfaction of knowing that you are doing your part. You might want to select a product from a line known to have contained RFID devices, like Gillette razors or Procter & Gamble makeup brands. If you're at a loss for what might be a good substitute, visit the Spychips.com website for suggestions. We also have information about products made by the spychippers, so you can see what items you might like to avoid the next time you shop.

Deactivate or remove any RFID chips on products you buy. If you absolutely must buy a product that contains a spychip, be sure to deactivate or remove the chip—preferably, prior to leaving the store.

Teach your children the value of privacy—both in words and through your actions. If you take steps to fight RFID, let them know. Your message will remain with them as a lasting reminder about how important it is to fight for privacy and civil liberties. Both of us keenly remember the boycott launched against Nestlé in response to its unethical methods for marketing baby formula to third-world mothers.[8] While we were very young at the time, our mothers made a point of telling us and explaining why we wouldn't be buying Nestlé products—something we remember vividly to this day.*

MODERATE STEPS

Following are some steps that take a bit more thought and effort but pay off richly in the fight against RFID. Consider doing one or more of these after you've mastered the smaller steps:

Write to companies that promote RFID and tell them how you feel. Companies are in existence because of consumers—not the other way around. (Though sometimes it can feel like we're not in control!) When companies and their boards realize that consumers object to their invasive practices, they'll listen—or they'll perish.

Write to your newspaper editor, or to your favorite blogger, commentator, radio host, or TV news anchor. Let them know that you object to RFID, and share a copy of this book with them.

---

*Back in the seventies, Nestlé hired women without any special training to pose as nurses and give away samples of baby formula to mothers in third-world countries. Because their milk dried up during the sample period, the mothers were forced to buy formula from Nestlé after the freebies were gone. Poor mothers tried to stretch the formula by mixing it with unsanitary water, resulting in thousands of infant deaths from disease and malnutrition. In response, a worldwide boycott was launched against the company, and the World Health Organization instituted ethics rules on the marketing of breast milk substitutes. Today, Nestlé is deeply involved with initiatives to tag its products with RFID.

Discuss the issue in your organization or company newsletter. Pick a single topic from this book, or do an overview. No matter what you write, it will be news to most of your colleagues.

Participate in an RFID protest. Visit the Spychips.com site to learn about planned protests. (If there isn't one planned in your area, you might want to skip ahead to **bold steps** to learn how to organize one yourself.)

Return unused products you've purchased from companies that promote RFID, and be sure to explain why you're returning the products. Why run the risk of taking home an unlabeled spychip? Your product return will send a powerful message to the spychippers—renounce RFID or lose your customers. In some cases, returning products can be even more effective than not buying them in the first place.

Educate your kids, family, friends, and coworkers about the dangers of RFID. Loan them this book—or better yet, buy them a book of their own.

Send us your tips about RFID devices you've found in consumer products, so we can tell others. But first, be sure that what you've found is not an EAS anti-theft device. You can usually tell by looking for the telltale microchip attached to an antenna that gives away a spychip. Or, if you enjoy watching executives squirm, try asking the company, "Hey! Is that a spychip in my underwear?"

Write to your state and federal lawmakers. We have a right to know when products contain tracking devices that could jeopardize our privacy or violate our principles. Tell politicians you want them to enact labeling legislation like our proposed *RFID Right to Know Act* that will allow us to make our own decisions about RFID. Need help contacting your representatives or crafting a letter? Visit Spychips.com for tips and links for reaching out. You can also refer lawmakers to the website, or better yet, send them a copy of this book so they can read the truth for themselves. But, while they can help us with labeling, remember not to depend on them to solve the problem. This fight is up to us!

Bold Steps

Organize a peaceful RFID protest. Let retailers and manufacturers see evidence of consumer displeasure with their plans. We'll even help you get the word out if you tell us about it in advance.

Wear an RFID protest T-shirt in public places and share the news about RFID with people who express interest in your message. You'd be amazed how interested people are in the topic if you give them a chance to talk about it. Be sure to do your homework first so you can answer questions that will inevitably arise if you wear a T-shirt in public.

Join us in tracking RFID patent developments. It's tough keeping up with the hundreds of RFID patent applications filed every year. It would be helpful if we could find some volunteers to follow developments and get the word out to the public. After all, eternal vigilance is the price of liberty.

Organize a speaking event. You can contact us directly for information on how we might help you motivate large audiences to get involved in fighting RFID. You can reach us at our website, Spychips.com.

Buy us billboard space. Depending upon where in the country you live, a monthly billboard rental starts at about one thousand dollars. If that's out of your price range, consider printing up a few flyers from our website and posting them on community bulletin boards at laundromats, supermarkets, college campuses, churches, and other sites that accept such postings.

Donate money. CASPIAN does accept gifts to fund its ongoing work. However, CASPIAN is not a 501c3 charitable organization, meaning that any gifts are not tax-deductible. The requirements for IRS tax-exempt status give the government the power to dictate and control an organization's activities. Obviously, being under the thumb of the federal government would run counter to both our mission and our philosophy. Since our founding in 1999, no one associated with CASPIAN has drawn a salary from their work with the organization. What we have accomplished to date, we have done through grassroots volunteers who've made individual gifts of time, effort, and money.

JOIN WITH US

RFID is a troubling technology with frightening implications. When people first learn what's coming, they can have one of several reactions. They can cover their ears and hum, denying the reality of the problem. They can admit the problem, but shrug their shoulders in resignation, believing there is nothing they can do to stop it. Or they can roll up their sleeves and take action.

If you choose the latter course, we want you to know that there is an entire community of concerned citizens just like you who have joined CASPIAN. We have thousands of members in over thirty countries around the world who oppose invasive retail strategies like loyalty cards, retail surveillance, and RFID. Members come from all positions on the political spectrum and all walks of life. To join CASPIAN, you simply tell us that you agree with this basic statement: "It is wrong to spy on people through the products and services they buy."

CASPIAN is a peaceful organization and does not advocate illegal or violent acts. We come together to support each other as we work to create positive changes in our communities and stores and spread news of a consumer revolution. We believe in using the power of the market—through our dollars and words—to achieve our ends.

You can learn more about joining CASPIAN and signing up for the free CASPIAN newsletter at our website: Spychips.com. There you'll also find up-to-date news about RFID and our ongoing work to keep spychips off consumer items.

Together We Can Change the Course of History

Though they seem huge and intimidating, corporations are actually quite simple. They behave in predictable ways. Nearly every action they take—including the adoption or rejection of privacy-invading technologies—is done to maximize profits. If their actions don't serve the mighty bottom line, they'll go out of business.

Since corporations depend on us, the customers, for their profits, we ultimately control what they do. Corporations are like puppets on a string. They dance to the tune of our dollars. Or you can think of them as plants growing towards the sunlight—they bend towards the money source that flows from our wallets. When we nourish them with our money, they grow in directions that are profitable for them. Conversely, if we withhold our shopping dollars from a company because of something it is doing, that action will become unprofitable and will quickly be dropped. Like a plant cut off from sunlight, the company will be forced to change direction until it reestablishes a beneficial relationship with consumers and their pocketbooks.

This is where the power of numbers comes into play. While each of us can make a difference working independently, when we pool our resources we can move mountains. By collectively spreading the word about RFID and withholding our money from businesses with a spychipping agenda, we can force seemingly unshakable global corporations to honor the wishes of the two-thirds of the population that objects to RFID.

Once the mega-businesses feel the economic impact of privacy invasion, they'll stop forcing their suppliers to adopt the technology. And, of course, once businesses find spychipping an unprofitable venture, the politicians who are supporting their agenda will fall in line. If we join together to take a stand against RFID, we can prevent spychips from invading our stores, homes, and bodies. Together we *can* change the direction of business, and, in so doing, we can change the course of history.

Things change at a breathtaking pace in the high-tech world of spychips. There have been many new developments since we first set this book to print. All are disquieting, as is the way with this technology, but we had a few "rolling on the floor" belly laughs, too.

Remember that taxpayer-funded relationship between North Dakota State University and Alien Technology? It's already begun to bear fruit—or rocks, that is. Their union has spawned RFID-enabled artificial rocks that listen.[1] They look like ordinary rocks, but they've been hollowed out and equipped with high-tech sensors capable of detecting oncoming footsteps from twenty to thirty feet away. Reportedly, the military is looking forward to dropping the devices from aircraft to sense approaching armies, but it doesn't take much imagination to see how that invention could be abused. One day soon, we may be taking a second look at the garden-variety rocks in our own backyards wondering if government-funded spychips have invaded the suburbs.

As we know, spychips have already invaded our nation's highways. To gather at bit more evidence, Liz headed to Houston to snap some last-minute photos and see the latest in highway surveillance in her home state of Texas. While there, she dropped in at an EZ-tag store and asked for a copy of their privacy policy. She spent half an hour cooling her heels in the lobby while staff combed their computers and made frantic phone calls. Eventually, they told

(Photo: Liz McIntyre)

This innocent-looking adhesive label contains a 3-inch-square RFID tag. It was found affixed to a Hewlett-Packard printer/scanner box purchased at a Texas Wal-Mart store in the fall of 2005. The only indication of the presence of the RFID tag is the 0.5-inch-square "EPC" symbol at the upper right.

her the privacy policy must be in storage—on microfiche—and that it would take some time to retrieve. Rather than grow cobwebs waiting, she left them our e-mail address. We're still waiting.

Here's a story that had us rolling on the floor. In a hilarious turn, IBM is offering its expertise as a privacy consultant.[2] (Isn't that a bit like having Enron executives teach accounting classes?) You heard right. The same company that applied for a patent on its notorious "person tracking unit" now claims it can help other companies adopt "privacy-optimized RFID solutions" and develop communication and education programs. Just what we all need—a little "re-education" from IBM. Attention potential IBM privacy clients: You'd do better to throw your money down a rat hole than rely on these guys for privacy advice. Once again, thanks for the laughter, IBM.

(PHOTO: LIZ MCINTYRE)

This Rafsec-brand RFID tag was embedded in the label shown above. Note the small computer chip in the center of the radiating antenna.

In fall of 2005, CASPIAN volunteers began finding spychipped electronics products in Wal-Mart stores around the country. We knew that Wal-Mart had planned to run a "trial" in the Dallas-Fort Worth area selling item-level tagged HP printers and scanners, but when one eagle-eyed volunteer, Nahsechay Oladipo, spotted a tiny half-inch EPC symbol on the HP proof of purchase label, we had confirmation.

When we looked deeper, we found that Wal-Mart stores had quietly begun selling HP printers packed in spychipped boxes—along with some Lexmark printers and Sanyo TV sets. Amazingly, Wal-Mart stores across the country are now sending RFID tags home with consumers without telling them anything about it. In fact, they're apparently not even informing their own workers, either. The employees we spoke to denied any knowledge of the tags, the chips, or the "EPC" symbol. "That's nothing at all," one Wal-Mart associate assured us. One manager in New Hampshire hadn't known RFID even existed until we showed him a tag affixed to a box on the shelf of his own store.

Since selling spychipped printer boxes was a clear violation of our call for

(Photo: Katherine Albrecht)

This Alien brand RFID tag was hidden under a label affixed to a Sanyo television box at a Dallas Wal-Mart. A Wal-Mart associate assured us "That's nothing at all."

a moratorium on item-level tagging, we responded with an "awareness raising event" outside of a Dallas Wal-Mart store that drew a huge crowd in protest. Two protests in New Hampshire quickly followed. Soon we were fielding inquires from journalists across the country, all wanting to know if their local Wal-Mart stores had crossed the item-level tagging line.

That was about the time our book began hitting the bookstores, setting off major tremors in the RFID world. Two days prior to its release, *Spychips* flew to the top of the Amazon bestseller charts, hitting number one as a "Mover and Shaker," making its way to the top-ten nonfiction bestseller list and spending weeks as a Current Events bestseller. Within its first four weeks alone, the book sold thousands of copies, and the journalistic and privacy communities called it "brilliantly written," "stunningly powerful," and "scathing." And in a nod to the book's focus on freedom, *Spychips* was awarded the prestigious Lysander Spooner Award for Advancing the Literature of Liberty and named "the best book on liberty" for 2005.

The RFID industry was shocked by the response to the book—and morti-

(PHOTO: COURTESY OF BILL BRYANT)

RFID opponents turn out in Dallas to protest Wal-Mart's use of item-level tagging.

fied at its contents. Being caught red-handed with elaborate people-tracking plans was embarrassing, to say the least. PR flacks and spinmeisters hurried to the defense of companies like IBM, NCR, Procter & Gamble, BellSouth, Philips, and Bank of America. But try as they might, they could find no errors in the book's pages—nor could they directly address the issues we raised.

We also caught the U.S. government with its own people-tracking plans. We had known the Department of Homeland Security was applying RFID tags to visitor documents at a handful of entry points. But Liz discovered a shocking Homeland Security Request for Information (RFI) seeking beefed-up RFID capabilities, like the ability to read RFID tags carried by passengers in speeding cars and to pinpoint pedestrians on the street from twenty-five feet away[3] What's more, the DHS indicated that this proposed RFID tag might be used for more than just visitor documents. We shuddered at the thought that a potent form of RFID might someday be incorporated into the federally standardized driver's licenses mandated by the Real ID Act. Spychipped licenses would give the government new powers of remote, silent surveillance over virtually all U.S. citizens as we move about in public.

Fortunately, national ID hit a raw nerve with the public, and across the country people began to fight back. One of the boldest challenges was launched from the tiny state of New Hampshire, when Representative Neal Kurk introduced a bill to prohibit New Hampshire from complying with the Real ID program.* Members of NH CASPIAN, the state branch of our privacy organization, rolled up their sleeves and got busy. Under the capable leadership of state director Joel Winters, we organized a coalition of over a dozen state political groups representing the left, the right, and everyone in between. The battle against the government's encroaching surveillance powers had begun.

We had industry representatives running scared. They made comical statements like, "We're scared to go to New Hampshire. They have gun racks on their motorcycles. They don't want anyone telling them what to do." For the record, Katherine (who calls New Hampshire home) has never seen a gun rack on a motorcycle, but she certainly agrees that the "Live Free or Die" state does not take kindly to being told what to do by the federal government—or the RFID industry. The bill sailed through the NH House, and we celebrated by holding a major rally outside the state capitol. But then the federal and industry arm-twisters descended on New Hampshire like a swarm of Gestapo. They threatened the state with penalties and reprisals, shook their fists, and banged the table. Ultimately, the NH state Senate rejected the original bill, and instead established a commission to report on the pros and cons of Real ID. The power mongers had once again overridden the will of the people.[4]

It was not the first time, either. Just weeks before, RFID and retail industry lobbyists had convinced the New Hampshire Senate to kill a "Right-to-Know" act that would have required a label on products containing RFID tags. Those same lobbyists have banded together to kill over a dozen labeling bills around the country in the past few years. Unless we voters start banding together ourselves, the industry will continue to use campaign contributions and polit-

---

*New Hampshire House Bill 1582: AN ACT prohibiting New Hampshire from participating in a national identification card system.

ical schmoozing to keep our elected officials in lobbyists' pockets, and hidden
RFID tags in *our* pockets—or even in our flesh.

The concerned citizens of Wisconsin see the threat posed by RFID
implants for human flesh, like the VeriChip, and have taken action. The state's
legislature passed a bill introduced by Representative Marlin D. Schneider that
makes it a crime to require implantation of a microchip into an individual.* [5]
The governor signed the legislation in 2006.[6] The threat of mandatory
implants became very real when a Cincinnati company called CityWatcher
had two of its employees injected with VeriChip implants to perform routine
job duties.[7] Other workers around the country worried that they, too, might
soon face pressure to "get chipped."

Former Secretary of Health and Human Services Tommy Thompson didn't
help matters when he left his Bush cabinet position in 2005 to join the board of
VeriChip.[8] Soon he was appearing on national TV touting RFID human
implants for all of us and encouraging citizens to get chipped as a way to link to
their medical records. He even suggested that the chips could replace military
dog tags, setting off alarm bells for our men and women in uniform. Despite his
high-sounding advice to the nation, as of this writing, Mr. Thompson has been
wise enough to avoid taking a chip implant himself. "He wants to see [the
VeriChip] in a real-world environment first," according to a VeriChip
spokesman. (That's just where we'd prefer not to see it, thank you very much.)

But the VeriChip people weren't laughing when researcher Jonathan
Westhues developed a device that could easily clone their supposedly secure
implant. Westhues showed how easily he could bump into someone with an
embedded VeriChip, silently capture the implant's number, and immediately begin emitting it himself in order to open locked doors or access
computer accounts.[9] [10] So much for security. But VeriChip's troubles were
only beginning.

The company's SEC Registration Statement—filed in anticipation of an

---

*Wisconsin Assembly Bill 290: prohibiting the required implanting of a microchip in an individ-
ual and providing a penalty.

(Photo: CASPIAN)

Katherine Abrecht spoke with former Secretary of Health and Human Services Tommy Thompson about the downsides of the VeriChip and gave him a copy of *Spychips* **in 2006 during his visit to New Hampshire.**

initial public offering of its stock, and its physician-chipping literature—are filled with damning factoids.[11] [12] Although the VeriChip is supposed to replace the MedicAlert bracelet to allow access to critical information in an emergency, its own documentation warns that the VeriChip might not be "merchantable or fit" for its purpose. What's more, VeriChip cautions that its patient database might not be available in an emergency, and that ambient radio waves in places like ambulances could prevent the chip from being read at all. Their solution? Move the patient away from ambulance interference to get a read on the implant. It's unpleasant to imagine paramedics carrying dying patients away from ambulances just to read their VeriChips. How likely is it that emergency officials will want to remove a heart attack patient from an ambulance in a thunderstorm or on an icy winter day? This sounds like a recipe for disaster.

Katherine met up with former Secretary Thompson when he recently visited New Hampshire. She gave him a copy of *Spychips* and asked for his views on the security downsides of the VeriChip. Mr. Thompson expressed surprise to learn of them. "I'm new to this whole VeriChip thing," he explained, and requested to meet at a future date to go over some of the issues.

During a promised follow-up phone call, Thompson's aide reported that the former DHS Secretary was reading *Spychips* with great interest—he asked his aide to read it, too. Let's hope the book helps cement his decision to remain "chip-free."

Human chipsters aren't the only ones worried about their products' security flaws. Chips on products could also be vulnerable to new hacking and virus techniques. A team of security researchers in the Netherlands managed to crack and decrypt Dutch passport chips (a perfect way for criminals to generate phony travel documents)[13] and another Dutch team showed that RFID systems could pick up viruses by reading tags containing malicious code.[14]

Then Australian security experts simulated a denial-of-service attack by overloading RFID readers with data, rendering them incapable of communicating. They concluded that the resulting system shutdown could spell disaster for "anyone trying to implement a[n] RFID system [with] mission critical or human life issues."[15] VeriChip-equipped emergency rooms, beware.

And in a final story, we've got word that it may be time to ditch your Dockers and lay off the Levi's. Levi Strauss & Co. confirmed it is testing item-level RFID-tagging, saying "a retail customer is testing RFID at one location [n the U.S.] . . . on a few of our larger-volume core men's Levi's jeans styles." However, Levi Strauss refused to name the location. "Out of respect for our customer's wishes, we are not going to discuss any specifics about their test," a Levi Strauss spokesperson said. The company also confirmed it is tagging Levi Strauss clothing products, including Dockers brand pants, at two of its franchise locations in Mexico.[16]

While Levi Strauss is reportedly using RFID "hang tags" that can be clipped from the garments at checkout, anyone who has read the evidence in this book knows external tags are just the opening volley in the spychipping game. We're going to be keeping a very close eye on this company and others that step over the line by tagging individual consumer items—especially clothing. Few things are more intimately connected with individuals than the clothes they wear.

We'll keep you posted on these and other breaking RFID developments at our website, Spychips.com. See you soon.

Katherine & Liz

Chapter 1 — Tracking Everything Everywhere

1. Rick Duris, "Just How Big Is RFID?" *Frontline Solutions*, 1 December 2003, available at www.frontlinetoday.com/frontline/article/articleDetail.jsp?id77382, accessed 15 April 2005.
2. C.P. Snow, *New York Times*, quote available at www.bartleby.com/63/36/3236.html, accessed 10 June 2005.
3. "State Senator Proposes Restrictions on RFID Spying," *San Francisco Business Times*, 24 February 2004, available at sanfrancisco.bizjournals.com/sanfrancisco/stories/2004/02/23/daily21.html, accessed 10 June 2005.
4. Kevin Reilly, "AMR Research Finds Wal–Mart Suppliers Spent Only Minimum Required to Comply with RFID Mandate," AMR Research press release, 20 December 2004, available at www.amrresearch.com/Content/View.asp?pmillid=17856&docid=12139, accessed 13 June 2005.
5. Greg Dixon of ScanSource, as quoted by Mark Riehl in "Partners Needed for RFID Success, Says ScanSource," *eChannelLine Daily News*, 9 August 2004, available at www.integratedmar.com/ecl–usa/story.cfm?item=18578, accessed 11 June 2005.

Chapter 2 — Spychips 101

1. MIT Auto–ID Center, "The New Network: Identify Any Object Anywhere Automatically," promotional brochure, MIT Auto–ID Center (Cambridge, MA).
2. "The Theremin Page," The Musical Museum (London, England) website, available at www.musicalmuseum.co.uk/theremin.html, accessed 9 February 2005.
3. Hon. Henry J. Hyde, "Introduction to 'Embassy Moscow: Attitudes and Errors,'" United States House of Representatives website, accessed 10 February 2005.
4. "Great Seal Exhibit," National Security Agency website, available at www.nsa.gov/museum/museu00029.cfm, accessed 28 January 2005.

5. Albert Glinksy, *Theremin: Ether Music and Espionage*. University of Illinois Press, 2000, pp. 273–274.
6. "The Theremin Page."
7. Raghu Das, "RFID Explained: An Introduction to RFID and Tagging Technologies," *ID TechEx* (Cambridge, UK), 2003.
8. Raghu Das, "RFID Explained . . ."
9. Hitachi, "World's smallest and thinnest 0.15 x 0.15 mm, 7.5 µm thick RFID IC chip." Hitachi website, available at http://www.hitachi.com/New/cnews/060206.html, accessed April 10, 2006.
10. Jonathan Collins, "Hitachi Unveils Integrated RFID Tag," *RFID Journal*, 4 September 2003, available at www.rfidjournal.com/article/articleview/556/1/1/.pdf, accessed 13 June 2005.
11. William Raymond Price, "Location of Lost Dentures Using RF Transponders," U.S. patent #6,734,795, assigned to D–TEC–DENT, filed on 9 August 2001, granted 11 May 2004.
12. "Near-Real Time Satellite Tags," Wildlife Conservation Society website, available at www.wcs.org/sw–around_the_globe/marine/marineafrica/greatwhitesharks/greatwhitesharksattag, accessed 18 February 2005.
13. "LLC and TBT Announce Partnership to Develop Printed Batteries," press release, Precisia website, available at www.precisia.net/news/precisia_news_20040506_01.html, accessed 7 February 2005.
14. Clive R. Van Heerden, "Fabric Antenna for Tags," U.S. patent #6677917, assigned to Koninklijke Philips Electronics NV, filed 25 February 2002, granted 13 January 2004.
15. "Flint Ink Revolutionizes Antennas," *RFID Journal*, 2005, available at www.rfidjournal.com/article/articleview/41/1/73/, accessed 6 June 2005.
16. Siemens, "Transforming Production with Tiny Transponders," Siemens website, available at w4.siemens.de/FuI/en/archiv/pof/heft2_02/artikel05/, accessed 5 March 2005.
17. Raghu Das, "RFID Explained . . ."

Chapter 3 — The Master Plan

1. Helen Duce, "Going Global," MIT Auto–ID Center, originally available at www.autoidcenter.org/CAM–AUTOID–EB–01.pdf, accessed 3 July 2003. This reference has been removed from the Auto–ID Center website and is now available at www.autoidlabs.com/whitepapers/CAM–AUTOID_EB–001.pdf.
2. European Union Data Protection Working Party, "Working Document on Data Protection Issues Related to RFID Technology," 19 January 2005, Article 29 Data Protection Working Party under Directive 95/46/EC of the European Parliament and of the Council of 24 October 1995 (Brussels, Belgium).
3. Mark Roberti, "Sponsors Guide," MIT Auto–ID Center, 24 June 2003.
4. David Brock, "The Compact Electronic Product Code: A 64–Bit Representation of the Electronic Product Code," white paper, 1 November 2001, MIT Auto–ID Center.
5. David Brock, "The Electronic Product Code (EPC): A Naming Scheme for Physical Objects," MIT Auto-ID Center (Cambridge, MA), 1 January 2001.

6. Rick Munarriz, "Interview with RFID Pioneer Kevin Ashton," The Motley Fool website, available at www.leighbureau.com/speakers/KAshton/essays/interview_fool.pdf, accessed 18 March 2005.

7. Joseph Jofish Kaye, "Counter Intelligence & Kitchen Sink White Paper," Massachusetts Institute of Technology website, available at xenia.media.mit.edu/~jofish/writing/kwsp.1.1.pdf, accessed 18 March 2005.

8. "VeriSign to Run EPC Directory," RFID Journal, 13 January 2004, available at www.rfidjournal.com/article/view/735, accessed 10 June 2005.

9. Constance Hays, "What They Know about You," New York Times, 14 November 2004, Section 3, p. 1.

10. Mark Roberti, "Sponsors Guide."

11. Robert Uhlig, "Anti–Theft Tags 'Pose Danger to Children,'" Telegraph Group Limited, 10 May 2001, available at portal.telegraph.co.uk/news/main.jhtml?xml=/news/2001/10/05/wmag05.xml&sSheet=/news/2001/10/05/ixhomef.html, accessed 11 June 2005.

12. "The New Network: Identify Any Object Anywhere Automatically," promotional brochure, MIT Auto–ID Center (Cambridge, MA).

13. John Stermer, "Radio Frequency ID: A New Era for Marketers?" Consumer Insight, 2001(Winter).

14. David Greenberg at the joint meeting of the Board of Overseers and Technology Board of the MIT Auto–ID Center, University Park Hotel (Cambridge, MA), 14 November 2002.

15. "Declaration—Utility or Design Patent Application," United States Patent and Trademark Office website, available at www.uspto.gov/web/forms/sb0001.pdf, accessed 13 June 2005.

16. "General Information Concerning Patents," United States Patent and Trademark Office website, available at www.uspto.gov/web/offices/pac/doc/general/index.html#whatpat, accessed 13 June 2005.

17. John R. Hind, James M. Mathewson, and Marcia L. Peters, "Identification and Tracking of Persons Using RFID–Tagged Items," U.S. patent application #20020165758, assigned to IBM, filed 2 November 2002.

18. Edwin Black, "IBM and the Holocaust," IBM and the Holocaust website, available at www.ibmandtheholocaust.com, accessed 28 February 2005.

Chapter 4 — The Spy in Your Shoe

1. John R. Hind, "Method to Address Security and Privacy Issues of the Use of RFID Systems to Track Consumer Products," U.S. patent application #20020116274, assigned to International Business Machines, filed 21 February 2001.

2. Jo Best, "US Clothes Firm Comes Clean on RFID Plans," Silicon.com website, 25 January 2005, available at www.silicon.com/research/specialreports/protectingid/0,3800002220,39127337,00.htm, accessed 13 June 2005.

3. Alok Jha, "Tesco Tests Spy Chip Technology," The Guardian, 19 July 2003, available at www.guardian.co.uk/uk_news/story/0,3604,1001211,00.html, accessed 14 June 2005.

4. Laurie Sullivan, "U.K. Retailer Goes on RFID Shopping Spree," *Information Week*, 17 January 2005, available at www.informationweek.com/show Article.jhtml?articleID=57701509, accessed 13 June 2005.

5. Elaine Allegrini, personal communication, on or around June 19, 2003.

6. Howard Wolinsky, "P&G, Wal–Mart Store Did Secret Test of RFID," *Chicago Sun–Times*, 9 November 2003, available at www.suntimes.com/output/lifestyles/cst–nws–spy09.html, accessed 14 June 2005.

7. Howard Wolinsky, "P&G, Wal–Mart Store Did Secret Test of RFID."

8. Silvio Albano, "Lessons Learned in the Real World," Powerpoint presented at the EPC Symposium held at McCormick Place Conference Center, Chicago, Illinois, 15–17 September 2003.

9. "Global Source Tagging: World Class Products, Services and Support," Sensormatic website, available at www.sensormatic.com/gst_www/support.asp, accessed 16 July 2005.

10. "Source Tagging: It All starts Here," Sensormatic pdf, available at www.sensormatic.com/GST_www/files/itallstartshere.pdf, accessed 30 March 2005.

11. "RFID-Enhanced EAS," Checkpoint Systems pdf, available at es.checkpointsystems.com/downloads/pdf/es107.pdf, accessed 16 July 2005.

12. "EAS: Success Stories: Hard Goods," Checkpoint Systems website, available at www.checkpointsystems.com/default.aspx?page=successstorieshardgoods, accessed 13 June 2005.

13. "North American Source Tagging Suppliers (By Category)," The Source Tagging Council via Wayback Machine at webarchive.org website, available at web.archive.org/web/20010202140500/http://synergy–stc.com/council/supplierlist.html, accessed 13 June 2005.

14. "Retailers Reap the Rewards of Source Tagging," Checkpoint Systems website, available at www.checkpointsystems.com/content/srctag/partic.aspx, accessed 7 October 2004.

15. "Chipless Smart Labels: Technology Evaluation," Smart Packaging Journal, *ID Tech Ex*, July/August 2002, pp. 8–10.

16. Jonathan Collins, "Checkpoint Buys 100 Million Tags," *RFID Journal*, 30 March 2004, available at www.rfidjournal.com/article/articleview/853/1/14/, accessed 14 June 2005.

17. "Checkpoint Systems Introduces EPC Solution Center," Checkpoint Systems press release, 12 January 2004, available at www.checkpointsystems.com/content/news/press_releases_archives_display.aspx?news_id=59, accessed 7 October 2004.

18. "Checkpoint Systems Demonstrates End–to–End Solutions for the Retail Consumer Product Supply Chain," Checkpoint Systems press release, November 2003, available at www.checkpointsystems.com/default.aspx?page=pressreleases archives&idnews=52, accessed 14 June 2005.

19. "Checkpoint Delivers Value–Added Solutions for Retailers around the World," in Checkpoint Systems *Retail News*, September 2004, p. 8. PDF available at www.checkpointsystems.com/docs/rn_9_1.pdf.

20. "Leading U.S. Drug Retailer to Roll Out Checkpoint's Digital RF EAS Chainwide," in Checkpoint Systems *Retail News,* September 2004, pp. 1, 4. PDF available at www.checkpointsystems.com/docs/rn_9_1.pdf.

21. "Food/Drug/Mass Merchandisers," Information Resources, Inc. website, available at www.infores.com/public/us/content/infoscan/fooddrugmass.htm, accessed 16 July 2005.

22. Harlan J. Onsrud, Jeff P. Johnson, and Xavier Lopez, "Protecting Personal Privacy in Using Geographic Information Systems," Photogrammetric Engineering and Remote Sensing, 1994, 60(9).

23. "Source Tagging Shoes Is a Step in the Right Direction," Checkpoint Systems *Retail News,* March 2005. PDF available at www.checkpointsystems.com/docs/ST_Shoes.pdf.

24. Clive R. Van Heerden, "Fabric Antenna for Tags," U.S. patent #6,677,917, assigned to Koninklijke Philips Electronics NV, filed on 25 February 2002, granted 13 January 2004.

Chapter 5 — There's a Target on Your Back

1. Charlie Schmidt, "Beyond the Bar Code," *MIT Technology Review,* March 2001, pp. 80–85.

2. Judith J. Leonard, "Great Ideas—Super Sleuth Supermarket Survey," Thomson South-Western website, available at www.swlearning.com/marketing/gitm/gitm11–5.html, accessed 9 March 2005.

3. "Frequently Asked Questions," Envirosell website, available at www.envirosell.com/europe/ee_faq.html, accessed 14 June 2005.

4. Craig Childress, "Table Tent Cards Are Finally Getting Some Respect," Nation's Restaurant News, 1996.

5. "Clients," Envirosell website, available at www.envirosell.com/clients/clients.html, accessed 14 June 2005.

6. Kenneth J. Chapman, et al., "Academic Integrity in the Business School Environment: I'll Get by with a Little Help from My Friends," *Journal of Marketing Education,* 2004, p. 236–249.

7. "Imagine . . . the House and the Store of the Future," originally available on P&G website at www.pg.com/champion/inside.jhtml?document=%2Fcontent%2Fen_US%2Fxml%2Fchampion%2Fret_inside_jun012000_hfuture.xml, Google cache accessed 26 November 2004.

8. "Inside P&G Brands: A Chip in the Shopping Cart," originally available on P&G website at www.pg.com/champion/inside.jhtml?document=%2Fcontent%2Fen_US%2Fxml%2Fchampion%2Fret_inside_jun012000_hfuture_repubblica.xml, Google cache accessed 26 November 2004.

9. "Imagine . . . the House and the Store of the Future,"

10. "Inside P&G Brands: A Chip in the Shopping Cart."
     Why show Coke drinkers a Pepsi ad? P&G explains: "As soon as a product finishes, its competitors automatically send their tempting and alternative ads." If people don't go for this plan on their own, P&G suggests getting them "to play this little game" by giving away TV sets to run the ads.

11. Ronald Gary Godsey, Marshall P. Haine, and Mary Elizabeth Scheid, "System and Methods for Tracking Consumers in a Store Environment," U.S. patent application #20020161651, assigned to The Procter & Gamble Company, filed 22 August 2001.

12. Mark Baard, *Wired News* reporter, personal e–mail communication with Katherine Albrecht, 16 June 2005 (among other sources).

13. Carrie Johnson, "Ahold Settles SEC Fraud Charges," *Washington Post*, 14 October 2004, available at www.washingtonpost.com/wp–dyn/articles/A30807–2004Oct13.html, accessed 15 June 2005.

14. Elizabeth Peroni, "Giant Food Store Revitalizes HBC at New Mechanicsburg Supermarket," Pennsylvania Food Merchants Association, June 2001, available at www.pfma.org/media/advisor/JUN01/retailer/Giant.html.

15. "About NCR—Overview," NCR Corporation website, available at www.ncr.com/about_ncr/aboutncr.htm, accessed 1 March 2005.

16. "About NCR—Overview."

17. "Data Center Availability: Facilities, Staffing, and Operations," Microsoft website, available at www.microsoft.com/resources/documentation/sql/2000/all/reskit/en–us/part4/c1461.mspx, accessed 15 June 2005.

18. Constance Hays, "What They Know about You," *New York Times*, 14 November 2004, section 3, p. 1.

19. Bio for Dan Odette, vice president Global Industry Consulting, Teradata, a division of NCR, available at www.teradata.com/t/go.aspx/page.html?id=112748, accessed 1 March 2005.

20. Jerome A. Otto and Dennis J. Seitz, "Automated Monitoring of Activity of Shoppers in a Market," U.S. patent #6,659,344, assigned to NCR Corporation, filed on 6 December 2000, granted 9 December 2003.

21. Werner Reinartz and V. Kumar, "Not All Customers Are Created Equal," Harvard Business School Working Knowledge, 29 July 2002, available at hbswk.hbs.edu/pubitem.jhtml?id=3028&t=customer, accessed 15 June 2005.

22. James E. Dion, "The Customer May Not Always Be Right: Customer Service Strategies for Survival Today," Powerpoint presentation, available at www.dion-co.com/downloads/cba2002customer.PPT, accessed 15 June 2005.

23. Marty Abrams, "Policy Practice: Double Edged Sword," *Direct*, 15 March 2001, available at www.directmag.com/mag/marketing_policy_practice/, accessed 15 June 2005.

24. Texas Instruments, "Customer Loyalty Mechanism with TI*RFID," originally available on Texas Instruments website at www.ti.com/tiris/docs/solutions/pos/loyalty.shtml, accessed 19 December 2003. This reference has been removed from TI's website and is now archived at web.archive.org/web/20040205161015/http://www.ti.com/tiris/docs/solutions/pos/loyalty.shtml.

25. "Arthur Blank & Co. Set for High–Volume RFID Card Production," Contactless News, 3 March 2005, available at www.contactlessnews.com/news/2005/03/03/arthur-blank-co-set-for-highvolume-rfid-card-production/, accessed 15 June 2005.

26. Beth Givens, "RFID and the Public Policy Void: Testimony to the California Legislature Joint Committee on Preparing California for the 21st Century, Senator Debra Bowen, Chair," Sacramento, CA, 18 August 2003.

27. John R. Hind, James M. Mathewson, and Marcia L. Peters, "Identification and Tracking of Persons Using RFID-Tagged Items," U.S. patent application #20020165758, assigned to IBM, filed 2 November 2002.

28. Vicki Ward, "Coming Everywhere Near You: RFID," IBM website, available at www.1.ibm.com/industries/financialservices/doc/content/landing/884118103.html, accessed 7 October 2004.

29. "RFID May Boost Service at Banks," *RFID Journal*, 25 April 2003, available at www.rfidjournal.com/article/articleview/396/1/1/, accessed 31 August 2003.

30. David D. Strunk, "System and Method for Interactive Advertising," U.S. patent #6,708,176, assigned to Bank of America, filed on 24 April 2003, granted 16 March 2005.

Chapter 6 — The RFID Retail Zoo

1. "Metro Opens High–Tech Shop and Claudia Approves," IBM press release, 28 April 2003, available at www–1.ibm.com/industries/wireless/doc/content/news/press release/872672104.html, accessed 1 October 2004.

2. "Future Store Shopping a Reality with IBM," IBM press release, 28 April 2003, available at www–1.ibm.com/industries/retail/doc/content/news/pressrelease/430887101.html, accessed 18 April 2005.

3. "Metro Opens High–Tech Shop and Claudia Approves."

4. "50 Ideas for Revolutionizing the Store through RFID," NCR Corporation, November 2003.

5. Ted Bridis, "Most Consumers Unaware of Online Pricing Tactics," AOL Business News, 1 June 2005, available at aolsvc.news.aol.com/business/article.adp?id=20050601000509990008, accessed 16 June 2005.

    Note: It seems other retailers haven't learned from the Amazon debacle. A recent Annenberg Public Policy Center study notes that other online retailers are still tailoring prices to customers. Perhaps these retailers are getting away with it for now because they can manipulate the prices secretly. The study indicates that the majority of consumers are not aware of the practice of customer–specific pricing.

    The Annenberg study reference is: Joseph Turow, Lauren Feldman, and Kimberly Meltzer, "Open to Exploitation: American Shoppers Online and Offline," Annenberg Public Policy Center of the University of Pennsylvania, June 2005, available at www.annenbergpublicpolicycenter.org/04_info_society/Turow_APPC_Report_WEB_FINAL.pdf.

6. Alan Glass, senior vice president electronic commerce, MasterCard International, testimony given before U.S. House Committee on Commerce, 30 April 1998.

7. Stephanie Simon, "Shopping with Big Brother: The Latest Trend in Market Research is Using Surveillance Devices Such as Hidden Microphones to Spy on

Shoppers," *Los Angeles Times*, 1 May 2002, available at www.latimes.com/templates/misc/printstory.jsp?slug=la–050102spy, accessed 1 May 2002.

8. Lorene Yue, "Not So Many Happy Returns at Some Stores," *Chicago Tribune*, 19 December 2004, available at www.chicagotribune.com/business/yourmoney/sns–yourmoney–1219onthemoney,0,7193391.story, accessed 7 March 2005.

9. Ariana Eunjung Cha, "Some Shoppers Find Fewer Happy Returns," *Washington Post*, 7 November 2004, available at www.washingtonpost.com/wp–dyn/articles/A30908–2004Nov6.html, accessed 7 March 2005.

10. Ariana Eunjung Cha, "Some Shoppers Find Fewer Happy Returns."

11. "Tokyo Cabs to Try RFID Payments," *RFID Journal*, 19 October 2004, available at www.rfidjournal.com/article/articleview/1197/1/1/, accessed 5 April 2005.

12. Per Olof Loof, "Complete Integrated Self–Checkout System and Method," U.S. patent #6,507,279, assigned to Sensormatic Electronics Corporation, filed 6 June 2001, granted 14 January 2003.

13. "Police Officer Fired for Smoking Tobacco," *Portsmouth Herald*, 22 June 2003, available at www.seacoastonline.com/2003news/06222003/south_of/35552.htm, accessed 19 April 2005.

14. King County Washington, "Focus on Employees: Healthy Incentives Benefits Program," King County website, available at www.metrokc.gov/employees/focus_on_employees/FAQ.aspx, accessed 19 April 2005.

15 "NRF to Re-create Metro Group's Future Store," Supermarket News, 17 November 2003, available at www.supermarketnews.com/xref.cfm?&ID=7840&xref=NRF, accessed 1 March 2005.

16. Rena Tangens, co-director of FoeBud in Bielefeld, Germany, personal communication, 2004.

Chapter 7 — Bringing It Home

1. Charlie Schmidt, "Beyond the Bar Code," *MIT Technology Review*, 2001, pp. 80–85.

2. "Speech Recognition Finds Its Voice," Accenture Technology Labs, 2004.

3. "Bringing the Reality of Aging Online," Accenture website, available at www.accenture.com/xd/xd.asp?it=enweb&xd=services%5Ctechnology%5Cresearch%5Cihs%5Creality_aging.xml, accessed 25 March 2005.

4. "Technology Comes Home," Accenture website, available at www.accenture.com/xdoc/en/services/technology/research/ihs/tech_ihs.pdf, accessed 21 February 2005.

5. "Technology Comes Home."

6. "Technology Comes Home."

7. "Technology Comes Home."

8. Dadong Wan, "Online Wardrobe," U.S. patent application #20020121980, assigned to Accenture, filed 2 March 2001.

9. "Technology Comes Home."

10. "Technology Comes Home."

11. Kenneth P. Fishkin and Jay Lundell, "RFID in Healthcare," draft, in *RFID:*

Applications, Security, and Privacy, S. Garfinkel and B. Rosenberg, editors, Addison–Wesley Professional, 2005.

12. Kenneth P. Fishkin and Jay Lundell, "RFID in Healthcare."

13. Larry J. Eshelman, et al., "Automatic System for Monitoring Independent Person Requiring Occasional Assistance," U.S. patent #6,611,206, assigned to Koninklijke Philips Electronics NV, filed on 15 March 2001, granted 26 August 2003.

14. Philips even boasts that its home monitoring system can listen for certain words in a person's speech to determine his or her mood at all times: "An example of set of rules for classifying an occupant as bored is . . . the occupant's sentences contain few words; a low incidence of words suggesting enthusiasm such as superlatives; quiet flat tone in the voice; lack of physical movement; little movement of the head or body; sighing sounds, etc." Once collected, "words indicative of mood may then be sent to the mental state/health status classifier for classification of the mood of the speaker." The system is so complete it can even measure "lack of eye contact with objects such as television or book in the scene." Source: Larry J. Eshelman, et al., "Automatic System for Monitoring . . ."

15. Carolyn Ramsey Catan, "Machine Readable Label Reader System for Articles with Changeable Status," U.S. patent #6,758,397, assigned to Koninklijke Philips Electronics NV, filed on 31 March 2001, granted 6 July 2004.

16. Carolyn Ramsey Catan, "Machine Readable Label Reader System . . ."

17. "Merloni Unveils RFID Appliances," *RFID Journal*, 4 April 2003, available at www.rfidjournal.com/article/view/369/1/1/, accessed 25 March 2005.

18. Michele Gershberg, "U.S. Advertisers Go Digital to Track Ads," Reuters, 18 August 2004, available at www.usatoday.com/tech/news/techinnovations/2004–08–18–rfid–plus–ads_x.htm, accessed 20 June 2005.

19. Charlie Schmidt, "Beyond the Bar Code."

20. Elizabeth Board, comments made at U.S. Chamber of Commerce event, "The Global Potential of Radio Frequency Identification," 14 June 2005.

21. If you still believe that food purchase monitoring will lead to "better food safety" and "faster product recalls," you should talk to Jill Crowson, the Seattle shopper who is suing the Kroger-owned QFC grocery chain for failure to alert shoppers to a recall. Soon after Ms. Crowson bought beef from QFC, it was recalled by the USDA for potential infection with mad cow disease. Though QFC pulled the remaining beef from its own shelves, the company made no effort to contact the loyalty card members who had purchased it, including Ms. Crowson, who fed the meat to her family.

   Source: Anita Ramasastry, "Do Stores that Offer Loyalty Cards Have a Duty to Notify Customers of Product Safety Recalls?" Findlaw's Writ, 5 August 2004, available at writ.news.findlaw.com/ramasastry/20040805.html, accessed 30 March 2005.

22. Charlie Schmidt, "Beyond the Bar Code."

23. "iceilings," Armstrong World Industries website, available at www.armstrong.com/commceilingsna/article7399.html, accessed 18 June 2005.

24. "RFID Applications," WiseTrack website, available at www.wisetrack.com/rfi-dapplications.pdf, accessed 22 March 2005.

25. Thomas Nello Giaccherini, "Inventory & Location System," U.S. patent application #20030214387, filed 20 May 2002.

26. Robert J. Orr and Gregory D. Abowd, "The Smart Floor . . ."

Chapter 8 — Talking Trash

1. Chris Lydgate and Nick Budnick, "Rubbish!" *Willamette Week*, 24 December 2002, available at www.wweek.com/story.php?story=3485&page=1#, accessed 23 February 2005.

2. Alona Wartofsky, "Star Dreck," *Washington Post*, 21 June 2004, available at www.washingtonpost.com/ac2/wp–dyn/A56765–2004Jun20?language=printer, accessed 28 February 2005.

3. "Guidelines on EPC for Consumer Products," EPCglobal website, available at www.epcglobalinc.org/public_policy/public_policy_guidelines.html, accessed 17 June 2005.

4. "Oregon Judge Recognizes Privacy Rights in Trash," TalkLeft, 1 December 2002, available at 64.233.187.104/search?q=cache:0EA4t0FGqNYJ:talkleft.com/new_archives/001114.html, accessed 28 March 2005.

5. *California v. Greenwood*, "The Supreme Court ruled that the Fourth Amendment does not prohibit the warrantless search and seizure of garbage left for collection outside the curtilage of a home," 16 May 1988, available at www.fightidentitytheft.com/shred_supreme_court.html.

6. John Preston, "Hollywood's Trash and Treasure," *Telegraph Magazine*, 9 October 2004, available at www.theage.com.au/articles/2004/10/07/1097089474917.html?from=storylhs&oneclick=true#mahrour–, accessed 28 February 2005.

7. John Preston, "Hollywood's Trash and Treasure."

8. Anne Schroeder, "Names & Faces," *Washington Post*, 19 June 2004, available at www.washingtonpost.com/wp–dyn/articles/A53722–2004Jun18.html, accessed 28 February 2005.

9. Ferrán Viladevall, "Por Su Basura les Conocereis," *El Mundo Suplementos Magazine*, 21 November 2004, available at www.el-mundo.es/magazine/2004/269/1100893833.html, accessed 28 February 2005.

10. "Muckrakers: Cash for Trash," BBC News, 27 July 2000, available at news.bbc.co.uk/1/hi/uk_politics/854047.stm, accessed 28 February 2005.

11. Hasan Suroor, "Scoops from Trash," *The Hindu*, 2 April 2001, available at www.hindu.com/2001/04/02/stories/0302000i.htm, accessed 28 February 2005.

12. Philip Wintour, "Economy Set to Meet Euro Test, Say MPs," *The Guardian*, 28 July 2000, available at www.guardian.co.uk/euro/story/0,11306,607306,00.html, accessed 28 February 2005.

13. Robert Barritz, "Method for Determining if a Publication Has Not Been Read," U.S. patent #6,600,419, assigned to Treetop Ventures LLC, filed on 31 January 2001, granted 29 July 2003.

14. Robert Barritz, "Method for Determining if a Publication Has Not Been Read."

15. Justine Kavanaugh, "Technology and Trash Team Up," *Government Technology*, February 1995, available at www.govtech.net/magazine/gt/1995/feb/trash.php, accessed 21 June 2005.

16. Theodore D. Geiszler, et al., "Reader System for Waste Bin Pickup Vehicles," U.S. patent #5,565,846, assigned to Indala Corporation, filed on 26 September 1994, granted 15 October 1996.

17. "Business Profile Bell South Investor Relations," BellSouth website, available at www.bellsouth.com/investor/ir_businessprofile.html, accessed 5 April 2005.

18. Barrett M. Kreiner and Donna K. Hodges, "System and Method for Utilizing RF Tags to Collect Data Concerning Post–Consumer Resources," U.S. patent application #20040129781, assigned to BellSouth, filed 8 July 2003.

19. Barrett M. Kreiner, et al., "Radio–Frequency Tags for Sorting Post-Consumption Items," U.S. patent application #20040133484, assigned to BellSouth, filed 8 January 2003.

20. "Breakthrough on 1–Cent RFID Tag," *RFID Journal*, 2 December 2002, available at www.rfidjournal.com/article/articleview/172/1/46/, accessed 25 March 2005.

21. Jonathan Collins, "RFID Fibers for Secure Applications," *RFID Journal*, 26 March 2004, available at www.rfidjournal.com/article/articleview/845/1/14/, accessed 1 February 2005.

22. Jim Rosenberg, "Printable Radio Tags Could Be Used to Track Newspapers," *Editor & Publisher*, 15 October 2003, available at www.mediainfo.com/editorandpublisher/headlines/article_display.jsp?vnu_content_id=2001992, accessed 15 October 2003.

23. Roberta M. McConochie and Jane Bailey, "Progress Toward Passive Measurement of Print," ESOMAR/ARF website, available at www.arbitron.com/downloads/ McConochieBaileyESOMAR2004.pdf, accessed 21 June 2005.

Chapter 9 — Yes, That's Your Medicine Cabinet Talking

1. "Truth Is Like the Sun," ThinkExist.com, available at en.thinkexist.com/quotation/truth_is_like_the_sun–you_can_shut_it_out_for_a/153597.html, accessed 22 June 2005.

2. "Elvis Week," Elvis Presley Enterprises website, available at www.elvis.com/graceland/calendar/elvis_week.asp, accessed 21 February 2005.

3. "The Med," Memphis Regional Medical Center website, available at www.the–med.org/themedhistory.pdf, accessed 19 February 2005.

4. Lea Nolan, et al., "An Assessment of the Safety Net in Memphis, Tennessee," The George Washington University Medical Center, March 2004.

5. Jonathan Collins, "Tracking Medical Emergencies," *RFID Journal*, 22 April 2004, available at www.rfidjournal.com/article/articleview/901/1/1/, accessed 19 February 2005.

6. Jonathan Collins, "Tracking Medical Emergencies."

7. "Alien Technology Corporation Successfully Completes RFID Trial at Memphis Medical Center," Alien Technology press release, 5 April 2004.

8. "Alien Technology Corporation Successfully Completes . . ."

9.   Sheila Dwyer, "Securing Patient Information," MedTech1.com, 17 May 2001, available at www.medtech1.com/success/device_stories.cfm/17/6, accessed 14 February 2005.

10. "Why RFID Is Critical," Precision Dynamics Corporation website, available at www.pdcorp.com/rfid/hc_why_rfid.html, accessed 8 November 2004.

11. "Why RFID Is Critical."

12. Dr. Lucian Leape, personal communication, 10 November 2004.

13. "HealthGrades Quality Study: Patient Safety in American Hospitals," *HealthGrades*, July 2004, p. 3.

14. "In-Hospital Deaths from Medical Errors at 195,000 Per Year, HealthGrades' Study Finds," *HealthGrades*, 27 July 2004, available at www.healthgrades.com/aboutus/ index.cfm?fuseaction=mod&modtype=content&modact=Media_PressRelease_Detail &&press_id=135, accessed 15 February 2005.

15. Jack DeAlmo, "RFID in the Pharmacy: Q&A with CVS; Q&A with Jack DeAlmo, VP Inventory Management and Merchandise Operations," in *RFID: Applications, Security, and Privacy*, S. Garfinkel, editor, Addison–Wesley Professional, 2005.

16. "EPC Auto ID in the Drug Channel," CVS via GMA website, available at www.gmabrands.com/events/docs/isld2004/epcauto.pdf, accessed 4 February 2005.

17. Dadong Wan, et al., "Online Medicine Cabinet," U.S. patent #6,539,281, assigned to Accenture Global Services GmbH, filed on 23 April 2001, granted 25 March 2003.

18. Ken Fishkin and Min Wang, "A Flexible, Low–Overhead Ubiquitous System for Medication Monitoring," Intel Research Seattle, University of Washington (Seattle), October 2003.

19. "RFID to Fight Counterfeiting of Viagra, Painkilling Drugs," Associated Press via *Information Week*, 15 November 2004, available at www.informationweek.com/ story/showArticle.jhtml?articleID=52601667, accessed 16 February 2005.

20. "FDA Announces New Initiative to Protect the US Drug Supply through the Use of Radiofrequency Identification Technology," press release, United States Food and Drug Administration, 15 November 2004, available at www.fda.gov/bbs/topics/news/2004/NEW01133.html, accessed 22 June 2005.

21. "VitalSense Integrated Physiological Monitoring System Brochure," Mini Mitter website, available at www.minimitter.com/Products/Brochures/ 900–0138–00_VS.pdf, accessed 14 February 2005.

22. Roland Piquepaille, "Using RFID Tags to Make Teeth," Roland Piquepaille's Technology Trends, 25 October 2004, available at radio.weblogs.com/0105910/ 2004/10/25.html, accessed 26 February 2005.

23. Nir Navor and Ronnie Botton, "Tampon Detection System," U.S. patent #6,348,640, filed on 20 March 2000, granted 19 February 2002.

24. Rosann Kaylor, "Healthcare Networks with Biosensors," U.S. patent application #20040078219, assigned to Kimberly–Clark Worldwide, Inc., filed 22 April 2004.

Chapter 10 — This Is a Stickup

1. "Real–World Showroom," Accenture website, available at www. accenture.com/xd/xd.asp?it=enweb&xd=services%5Ctechnology%5Ctech_ rwshowroom.xml, accessed 21 June 2005.
2. "Real–World Showroom."
3. "About Accenture," Accenture website, available at www.accenture.com/ xd/xd.asp?it=enweb&xd=aboutus\about_home.xml, accessed 21 February 2005.
4. "Clients," Accenture website, available at www.accenture.com/xd/xd.asp?it= enweb&xd=services%5Csba%5Csba_who_maclient.xml, accessed 28 February 2005.
5. Cindy Southworth, personal communication, 9 March 2005.
6. "Stalking," U.S. Department of Justice Office of Community Oriented Policing Services, 5 January 2004.
7. "First Comprehensive Review of Stalking in UK Published by Chubb Insurance," PR Newswire on behalf of Chubb Insurance Europe, 7 May 2004, available at www.prnewswire.co.uk/cgi/release?id=122401, accessed 26 February 2005.
8. "Stalking."
9. David Teather, "Man Arrested over GPS 'Stalking,'" *The Guardian*, 6 September 2004, available at www.guardian.co.uk/usa/story/0,12271,1297892,00.html, accessed 26 February 2005.
10. Barbara Ross and Tracy Connor, "Say Grocery Guy Delivered 'Terror,'" *New York Daily News*, 8 October 2004, available at nydailynews.com/front/story/ 240064p–205757c.html and www.nydailynews.com/front/story/240064p– 205757c.html, accessed 26 February 2005.
11. David Sorkin, "RF Tracker Request for Data," RF Tracker website, available at www.rftracker.com/info.html, accessed 26 February 2005.
12. "Female Employee Finds Web Cam under Her Desk," WFTV Florida Channel 9, 20 May 2004, available at www.wftv.com/news/3328543/detail.html, accessed 7 February 2005.
13. "Navy Finds Video Camera in Female Sailors' Shower," Associated Press, 4 May 2004, available at www.buckspipe.com/modules/news/article.php?item_ id=331, accessed 7 February 2005.
14. "Casino Fined for Hidden Cameras' Wandering Eyes," *USA Today*, 16 December 2004, available at www.usatoday.com/travel/hotels/2004-12-16-casino- cameras_x.htm, accessed 19 February 2005.
15. Robert O'Harrow, *No Place to Hide: Behind the Scenes of Our Emerging Surveillance Society*, Free Press, 2005, p. 179.
16. "Legal Loopholes Protect Video Voyeurs," CNN.com, 8 February 2005, available at www.cnn.com/2005/LAW/02/08/video.voyeur.ap/, accessed 9 March 2005.
17. Einat Amitay and Aya Soffer, "Personal Index of Items in Physical Proximity to a User," U.S. patent application #20050067492, assigned to IBM, filed 30 September 2003.
18. "Design against Crime," Crime Reduction website, available at www.crimere- duction.co.uk/securedesign14.htm, accessed 22 June 2005.

19. Chris Adams, "The Chipping of Goods Initiative," Police Scientific Development Branch (Hertfordshire, UK), October 2004.
20. "The Booster Bag Problem," Alert website, available at www.alertmetalguard.com/Default.asp?ID=6, accessed 8 March 2005.
21. "Hard-to-Shop-for People on Your Holiday List? How about an Electronic Wallet for Their Wrists?" ExxonMobil website, available at www2.exxonmobil.com/Corporate/Newsroom/Newsreleases/xom_nr_041202., accessed 20 May 2005.
22. John Schwartz, "Students Find Hole in Car Security Systems," *New York Times*, 28 January 2005, available at www.newyorktimes.com/2005/01/28/science/28cnd–key.html?ei=5094&en=48eb306a45a3b7aO&hp=&ex=1106974800&oref=login&partner=homepage&pagewanted=print&position, accessed 29 January 2005.
23. "Test of Air Force Radio Jams Area's Garage Door Openers," *USA Today*, 20 May 2004, available at www.usatoday.com/news/offbeat/2004-05-20-doors-jammed_x.htm, accessed 23 June 2005.
24. Mark Willoughby, "Securing RFID Information," *Computerworld*, 20 December 2004, available at www.computerworld.com/printthis/2004/0,4814,96051,00.html, accessed 9 March 2005.
25. Claire Swedberg, "Congress Considers Evacuation Tracking," *RFID Journal*, 7 February 2005, available at www.rfidjournal.com/article/articleview/1392/1/1/, accessed 9 March 2005.
26. John Gilmore, "RFID and Assassins," personal communication, 10 March 2005.
27. Matthew Wald, "New High-Tech Passports Raise Snooping Concerns," *New York Times*, 29 November 2004, available at www.wired.com/news/privacy/0,1848,66686,00.html, accessed 26 February 2005.
28. "Electronic Passport, Proposed Rule, 22 CFR Part 51, Public Notice 4993, RIN 1400-AB93," U.S. Department of State, 18 February 2005.
29. Matthew Wald, "New High–Tech Passports Raise Snooping Concerns."

Chapter 11 — Downshifting into Surveillance Mode

1. "The Privacy Bulletin," 1990, Special Issue, August, Volume 6, Number 2 (Sydney Australia). As cited in: Sheri Alpert and Kingsley E. Haynes, "Privacy and the Intersection of Geographical Information and Intelligent Transportation Systems," available at www.spatial.maine.edu/tempe/alpert.html, accessed 24 November 2004.
2. "Houston TranStar Fact Sheet 2003," Houston TranStar website, available at www.houstontranstar.org/about_transtar/docs/2003_fact_sheet_1.pdf, accessed 25 February 2005.
3. "Automated Vehicle Identification," Houston TranStar website, available at www.houstontranstar.org/about_transtar/docs/2003_fact_sheet_2.pdf, accessed 25 February 2005.
4. "Automated Vehicle Identification."
5. Matthew Amorello, "Fast Lane Program," Massachusetts Turnpike Authority website, available at www.massturnpike.com/travel/fastlane/, accessed 20 June 2005.
6. Shamus Toomey, "No I–Pass? Prepare To Pay Double," *Chicago Sun–Times*, 26

August 2004, available at www.suntimes.com/output/news/cst–nws–toll26.html, accessed 6 September 2004.

7. Fred Philipson, "Fast Lane," *Government Technology*, September 2004, available at www.govtech.net/magazine/story.php?id=91366&issue=9:2004, accessed 23 March 2005.

8. Road Guy, "Is Big Brother in the Tollbooth," *Daily Press*, reprinted at ITS America website, available at www.itsa.org/ITSNEWS.NSF/0/93a8b5 bfdeff96da85256c71004f51d4?OpenDocument, accessed 4 October 2003.

9. Jerry Werner, "More Details Emerge about the VII Effort," National Associations Working Group for ITS, 15 May 2004, available at www.ntoctalks.com/icdn/vii_details_itsa04.html, accessed 26 February 2005.

10. "Early Alert: ABI Research Flags Large DSRC Market Later This Decade," ABI Research press release, available at www.abiresearch.com/abiprdisplay.jsp?pres-sid=172, accessed 15 June 2005.

11. Jeffrey F. Paniati, "Vehicle Infrastructure Integration," Federal Highway Administration, U.S. Department of Transportation, available at www.itsa.org/resources.nsf/Files/VII_PM_01_Paniati_What_Is_VII/$file/VII_ PM_01_Paniati_What_Is_VII.pdf, accessed 26 February 2005.

12. Ralph Robinson, "VII Use Cases," Ford Motor Company, available at www.itsa.org/resources.nsf/Files/VII_PM_05_Robinson_Use_Cases/$file/ VII_PM_05_Robinson_Use_Cases.pdf, accessed 26 February 2005.

13. Jerry Werner, "USDOT Outlines the New VII Initiative at the 2004 TRB Annual Meeting," National Transportation Operations Coalition, 27 January 2004, available at www.ntoctalks.com/icdn/vii_trb04.php, accessed 26 February 2005.

14. Jerry Werner, "More Details Emerge about the VII Effort," National Associations Working Group for ITS, 15 May 2004, available at www.ntoctalks.com/icdn/vii_details_itsa04.html, accessed 26 February 2004.

15. Umar Riaz, et al., "Why Telematics Is Moving into the Realm of Transforming Technologies," Accenture website, available at www.accenture.com/xd/ xd.asp?it=enweb&xd=ideas%5Coutlook%5Cpov%5Cpov_telematics.xml, ac-cessed 10 December 2004.

16. Andrea Estes, "Fee Eyed for Those Who Drive into Hub," *Boston Globe*, 30 March 2005, available at www.boston.com/news/local/articles/2005/ 03/30/fee_eyed_for_those_who_drive_into_hub/, accessed 5 April 2005.

17. Robert Salladay, "DMV Chief Back Tax by Mile," *Los Angeles Times*, 16 November 2004, available at www.latimes.com/news/local/politics/cal/ la–me–dmv16nov16,0,987891.story?coll=la–news–politics–california, accessed 5 April 2005.

18. Michael Freitas, "VII Applications," ITS Joint Program Office, U.S. Department of Transportation, a.org/resources.nsf/Files/VII_PM_07_Freitas_Fed_Apps/$file/ VII_PM_07_Freitas_Fed_Apps.pdf, accessed 26 February 2005.

19. Jonathan Collins, "Automotive RFID Gets Rolling," *RFID Journal*, 13 April 2004, available at www.rfidjournal.com/article/articleview/866/1/1/, accessed 18 June 2005.

20. Randy Roebuck, "DSRC Technology and the DSRC Industry Consortium (DIC) Prototype Team," prepared by SIRIT Technologies for ARINDC/U.S. DOT, 28 January 2005.

21. Greg Lucas, "DMV Information Sold Illegally, State Audit Finds Agency Also Reaped Profits by Overcharging Clients," *San Francisco Chronicle*, p. A19.

22. Texas House Bill "H.B. No. 2893," proposed amendment to *Chapter 601, Transporation Code*, 2005.

23. "eGo Windshield Sticker Tag," TransCore website, available at www.transcore.com/product_profiles/411468.pdf, accessed 20 June 2005.

24. "Electronic Vehicle Registration (EVR)," TransCore website, available at www.transcore.com/markets/pdf/EVR%20Application%20Profile_ITSA03.pdf, accessed 14 June 2005.

25. "Gartner Reports Strong Opposition to a U.S. National Identity Program," Gartner press release, 12 March 2002, available at www.gartner.com/5_about/press_releases/2002_03/pr20020312a.jsp, accessed 20 June 2005.

26. Declan McCullagh, "House Approves Electronic ID Cards," CNET News.com, 10 February 2005, available at news.com.com/House+approves+electronic+ID+cards/2100-1028_3-5571898.html, accessed 23 February 2005.

Chapter 12 — The Chips that Won't Die

1. James M. Mathewson II and Marcia L. Stockton, "Using Radio Frequency Identification with Transaction–Specific Correlator Values Written on Transaction Receipts to Detect and/or Prevent Theft and Shoplifting," U.S. patent application #20050073417, assigned to IBM, filed 19 September 2003.

2. Katherine Albrecht, Liz McIntyre, and Beth Givens, "Position Statement on the Use of RFID on Consumer Products," CASPIAN and Privacy Rights Clearinghouse, 14 November 2003, available at www.spychips.com/jointrfid_position_paper.html and www.privacyrights.org/ar/RFIDposition.htm, accessed 14 November 2003.

3. Richard Shim, "RFID Blocker Tags Developed," Silicon.com, 28 August 2003, available at www.silicon.com/software/applications/0,39024653,10005771,00.htm, accessed 28 February 2005.

Chapter 13 — Adapt or Die

1. Helen Duce, "Message Development," MIT Auto–ID Center, originally available on MIT Auto-ID Center website at www.autoidcenter.com/media/communications.pdf, accessed 5 July 2003. This reference has been removed from the Auto-ID Center website and is now mirrored at cryptome.org/rfid/communications.pdf.

2. Helen Duce, "Public Policy: Understanding Public Opinion," MIT Auto–ID Center, originally available on MIT Auto–ID Center website at www.autoidcenter.com/publishedresearch/cam–autoid–eb002.pdf, accessed 5 July 2003. This reference has been removed from the Auto–ID Center website and is now mirrored at cryptome.org/rfid/cam–autoid–eb002.pdf.

3. Helen Duce, "Message Development."

4. "Managing External Communications," Fleishman–Hillard for the MIT Auto-ID Center, originally available at www.autoidcenter.com/media/external_comm.pdf, accessed 5 July 2003. This reference has been removed from the Auto-ID Center website and is now mirrored at cryptome.org/rfid/external_comm.pdf.

5. "Auto-ID Center Q & A," MIT Auto–ID Center, originally available at www.autoidcenter.com/new_media/media_kit/questions_answers.pdf, accessed 7 July 2003. This reference has been removed from the Auto–ID Center website and is now mirrored at cryptome.org/rfid/questions_answers.pdf.

6. Phyllis L. Kim, "Auto-ID Center Communications," Fleishman–Hillard for the MIT Auto-ID Center, originally available at www.autoidcenter.org/media/pk–fh.pdf, accessed 5 July 2003. This reference has been removed from the Auto-ID Center website and is now mirrored at cryptome.org/rfid/pk–fh.pdf.

7. Helen Duce, "Message Development."

8. Helen Duce, "Message Development."

9. Helen Duce, "Public Policy: Understanding Public Opinion."

10. Helen Duce, "Public Policy: Understanding Public Opinion."

11. Helen Duce, "Message Development."

12. Helen Duce, "Message Development."

13. Rick Munarriz, "Interview with RFID Pioneer Kevin Ashton," The Motley Fool interview reproduced by Leigh Bureau, 15 December 2004, available at www.leigh-bureau.com/speakers/KAshton/essays/interview_fool.pdf, accessed 12 May 2005.

14. "Managing External Communications."

15. Mark Baard, "RFID Cards Get Spin Treatment," Wired News, 29 March 2005, available at www.wired.com/news/privacy/0,1848,67025,00.html, accessed 22 June 2005.

16. Helen Duce, "Public Policy: Understanding Public Opinion."

17. Elizabeth Board, personal communication, 18 April 2005.

18. John Rabun, For Healthcare Professionals: Guidelines on Prevention of and Response to Infant Abductions, Alexandria: National Center for Missing & Exploited Children, 2003, p. 92.

19. John Rabun, For Healthcare Professionals.

20. Ari Juels, Ronald L. Rivest, and Michael Szydlo, "The Blocker Tag: Selective Blocking of RFID Tags for Consumer Privacy," available at www.rsasecurity.com/rsalabs/staff/bios/ajuels/publications/blocker/blocker.pdf, accessed 22 April 2005.

21. "Inside P&G Brands: A Chip in the Shopping Cart," originally available on P&G website at www.pg.com/champion/inside.jhtml?document=%2Fcontent%2Fen_US%2Fxml%2Fchampion%2Fret_inside_jun012000_hfuture_repubblica.xml, Google cache accessed 26 November 2004.

22. John R. Hind, James M. Mathewson, and Marcia L. Peters, "Identification and Tracking of Persons Using RFID–Tagged Items," U.S. patent application #20020165758, assigned to IBM, filed 2 November 2002.

23. Mark Baard, "Errant E–mail Shames RFID Backer," Wired News, 12 January

2004, available at www.wired.com/news/privacy/0,1848,61868,00.html?
tw=wn_story_top5, accessed 24 June 2005.

24.   Mark Baard, "Errant E–mail Shames RFID Backer."

25.   Andy McCue, "Digital Blunder Expose 'Dirty Tricks' in RFID War,"
       Silicon.com/CNet networks, 12 January 2004, available at www.silicon.
       com/hardware/storage/0,39024649,39117735,00.htm, accessed 22 June 2005.

26.   "Grocery Manufacturers Apologise to Anti–RFID Activist Over Slur," *Sydney
       Morning Herald* online, 13 January 2004, available at www.smh.com.au/
       articles/2004/01/13/1073877812378.html, accessed 22 June 2005.

27.   Abby Dinham, "IBM Hits Back at RFID Critics," ZDNet, 29 April 2004, available
       at www.zdnet.com.au/news/business/0,39023166,39146173,00.htm, accessed 8
       April 2005.

28.   John R. Hind, et al., "Identification and Tracking of Persons Using RFID-
       Tagged Items."

29.   Graeme Wearden, "How to Get Consumers to Swallow Electronic Tags," ZDNet,
       18 October 2004, available at news.zdnet.co.uk/hardware/emeringtech/
       0,39020357,39170565,00.htm, accessed 2 April 2005.

30.   "RFID and Consumers: Understanding Their Mindset," CapGemini and
       National Retail Federation, 2004, available at www.nrf.com/download/
       NewRFID_NRF.pdf, accessed 25 March 2005.

31.   Jonathan Collins, "Consumers More RFID–Aware, Still Wary," *RFID Journal*, 8
       April 2005, available at www.rfidjournal.com/article/articleview/1491/1/1/,
       accessed 10 April 2005.

32.   "Managing External Communications."

Chapter 14 — Are You Next?

1.   Rob Stein, "Bar Code Implant Calls Up Medical Data, FDA Approval Draws Fire
       from Advocates of Personal Privacy," *Washington Post*, reprinted at *San
       Francisco Chronicle*, 14 October 2004, available at sfgate.com/cgi–bin/
       article/article?f=/c/a/2004/10/14/MNGQA99FDM1.DTL, accessed 27 June
       2005.

2.   "Cuban American Bar Assn v. Christopher: Complaint," United States District
       Court, Southern District of Florida, 1994.

3.   "Cuban American Bar Assn v. Christopher: Complaint."

4.   "Cuban American Bar Assn v. Christopher: Complaint."

5.   Lynne Brakeman, contributing editor, "New DoD System Tracks Refugees,"
       Automatic ID News (Advanstar Communications Inc.), Volume 10, No. 13
       (December 1994), pp. 14–17.

6.   Lynne Brakeman, "New DoD System Tracks Refugees."

7.   Nat Fahy, "Marine Lieutenant Runs Modern Day 'Ellis Island'," Navy Office of
       Information, available at www.chinfo.navy.mil/navpalib/news/mcnews/
       mcn95/mcn95033.txt, accessed 28 November 2003.

8.   John Penido, fire chief (San Marino, California), "Fire Service Response to a
       Biological Event," *Bioterrorism: Homeland Defense: The Next Steps*, conference,
       TRANS-ATTACK PANEL, proceedings available at www.rand.org/nsrd/
       bioterr/pdf/cp–JPenido.pdf, accessed 16 July 2005.

9. Sheila Mitchell, "Global Supply Chain, RFID & GTN Standards Conference, 14th October, in Toronto," promotional e–mail received by John Young from Softmatch, archived at cryptome.org/rfid–fun–spam.htm, accessed June 26, 2005.
10. Audrey Hudson, "Bug Devices Track Officials at Summit," *Washington Times*, 14 December 2003, available at washingtontimes.com/national/20031214–011754–1280r.htm, accessed 26 June 2005.
11. Audrey Hudson and Betsy Pisik, "Summit Group Confirms Use of ID Chip.pdf," *Washington Times*, 17 December 2003, available at washingtontimes.com/national/20031217–115051–5373r.htm, accessed 26 June 2005.
12. "U.S. Military Clarifies RFID Mandate," *RFID Journal*, 10 October 2003, available at www.rfidjournal.com/article/articleview/608/1/1/, accessed 24 June 2005.
13. "RFID Technology in Iraq," Precision Dynamics Corporation website, 20 May 2003, available at www.pdcorp.com/healthcare/rfid_militaryuse.html, accessed 24 June 2005.
14. "Smart Military Medical 'Dog Tags'," Pacific Northwest National Laboratory website, available at www.technet.pnl.gov/sensors/electronics/projects/ES4rfT–DogTag.stm, accessed 10 June 2005.
15. "RFID Tracked Casualties in Iraq," *RFID Journal*, 19 May 2003, available at www.rfidjournal.com/article/articleview/431/1/44/, accessed 26 June 2005.
16. "2000 Census Data—Age and Sex Profile for United States," Annie E. Casey Foundation website, available at www.aecf.org/cgi–bin/aeccensus.cgi?action=profileresults&area=00N&printerfriendly=0&section=1, accessed 3 May 2005.
17. "Enterprise Charter School Applies Texas Instruments' RFID Contactless Technology for Multiple Applications," Texas Instruments website, 2003, available at www.ti.com/tiris/docs/news/_news_releases/2003/rell9–15–03.shtml, accessed 27 June 2005.
18. "Report to the Governor, the Temporary President of the Senate, and the Speaker of the Assembly on the Educational Effectiveness of the Charter School Approach in New York State," the State Education Department, University of the State of New York, 5 December 2003, p. 13.
19. Ann Bednarz, "RFID Everywhere: From Amusement Parks to Blood Supplies," NetworkWorld, 3 May 2004, available at www.networkworld.com/news/2004/0503widernetrfid.html?page=2, accessed 24 June 2005.
20. Matt Richtel, "In Texas, 28,000 Students Test E–Tagging System," CNET News.com, 17 November 2004, available at news.com.com/In+Texas,+28,000+students+test+an+electronic+eye/2100–1039_3–5456061.html and news.com.com/In+Texas%2C+28%2C000+students+test+e–tagging+system+–+page+2/2100–1039_3–5456061–2.html?tag=st.next, accessed 26 June 2005.
21. "The Board of Trustees," Spring Independent School District website, available at www.springisd.org/default.aspx?name=admin.board, accessed 6 May 2005.
22. Matt Richtel, "A Student ID That Can Also Take Roll," *New York Times*, 17 November 2004, available at richtel-a-student-id-that-can-also-take-roll, accessed 24 June 2005.

23. "Project Summary: ITR/SII+IM+EWF: Technologies for Sensor–Based Wireless Networks of Toys for Smart Developmental Problem–Solving Environments," UCLA Smart Kindergarten, available at nesl.ee.ucla.edu/projects/smartkg/docs/proposal.htm, accessed 24 June 2005.

24. Chris Sutton, "Wired Classroom Gives Educators Insight into Child Learning," *UCLA Engineer*, 11 June 2003.

25. "Project Summary: ITR/SII+IM+EWF . . ."

26. "News in Brief from Northern California," Associated Press, as reprinted in the *San Jose Mercury News*, 28 January 2005, available at www.mercurynews.com/mld/mercurynews/news/local/states/california/northern_california/10759096.htm, accessed 24 June 2005.

27. "California Company Pulls out of Program to Track Student Movements," ACLU, 16 February 2005, available at www.aclu.org/StudentsRights/StudentsRights.cfm?ID=17524&c=161, accessed 24 June 2005.

28. "Woodward Laboratories Announces the Release of iHygiene," Woodward Laboratories pdf, available at www.woodwardlabs.com/pdfs/iHygiene_Press_Release.pdf, accessed 5 January 2005.

29. "Member Firms Description: Cincinnati/Northern KY Firms," Executive Women International website, available at www.ewicincinky.com/firmdescription.htm, accessed 5 May 2005.

30. "Member Spotlight," Midland Hispanic Chamber of Commerce website, available at www.midlandhcc.com/mhcc/?section=2, accessed 5 May 2005.

31. Ameripride homepage, Ameripride Services Inc. website, available at www.ameripride.com, accessed 5 May 2005.

32. "Star City Casino: Silent Commerce Chips Away at Casino Wardrobe Worries," Accenture website, available at www.accenture.com/xd/xd.asp?it=enweb&xd=services%5Ctechnology%5Cvision%5Cstar_city_casino.xml, accessed 2 June 2005.

33. "Verichip RFID Implants in Mexican Attorney General's Office Overstated," Spychips website, 29 November 2004, available at www.spychips.com/press–releases/mexican–implant–correction.html, accessed 24 June 2005.

34. Will Weissert, "Microchips Implanted in Mexican Officials," MSNBC.com, 14 July 2004, available at www.msnbc.msn.com/id/5439055/, accessed 26 June 2005.

35. Jonathan Kent, "Malaysia Car Thieves Steal Finger," BBC News, 31 March 2005, available at news.bbc.co.uk/2/hi/asia–pacific/4396831.stm, accessed 5 May 2005.

36. "Baja Beach Club in Barcelona," SenorStag.com, available at www.senorstag.com/index/713, accessed 27 April 2005.

37. Jonathan Watts, "Experts from Around the World Join Largest Ever Forensic Investigation," *The Guardian*, 4 January 2005, available at www.guardian.co.uk/tsunami/story/0,15671,1382758,00.html, accessed 27 June 2005.

38. Bill Christensen, "Chip Implants Proposed to Halt Blackmarket Cadaver Trade," *Live Science*, 15 February 2005, available at www.livescience.com/scienceoffiction/technovel_organs_050215.html, accessed 27 June 2005.

39. Rob Stein, "Bar Code Implant Calls Up Medical Data . . ."

40. "CASPIAN Special Report: FDA Letter Raises Questions about VeriChip Safety, Data Security," Spychips website, 19 October 2004, available at www.spychips.com/reports/verichip–fda.html, accessed 27 June 2005.

41. "A Primer on Medical Device Interactions with Magnetic Resonance Imaging Systems," U.S. Food and Drug Administration, 7 February 1997, available at www.fda.gov/cdrh/ode/primerf6.html, accessed 27 June 2005.

42. See www.solusat.com.mx.

43. "Vinoble to Offer RFID Personal Mobile Location Technology Service." Vinoble, Inc., press release, available at finance.lycos.com/qc/news/ story.aspx?symbols=BB:VNBL&story=200504230705_CCN_0423001n.

44. Peter Seth Edelstein and Benjamin Theodore Nordell II, "Method and Apparatus for Locating and Tracking Persons," U.S. patent application #20040174258, filed 29 August 2003.

45. George M. Vodin, "Method and Apparatus for Remote Monitoring and Control of a Target Group," U.S. patent application #20030071734, filed 23 September 2002.

Chapter 15 — Your Tax Dollars at Work

1. QuoteWorld website, available at www.quoteworld.org/author.php? thetext=Ronald+Wilson+Reagan+(b.+1911), accessed 26 June 2005.

2. "My Doomsday Weapon: An Exhibition by Jakob S. Boeskov," The Thing website, available at www.backfire.dk/JB/index.html, accessed 10 May 2005.

3. "Data Mining: Federal Efforts Cover a Wide Range of Uses," United States General Accounting Office (Washington), report number GAO–040548, May 2004.

4. "Data Mining: Federal Efforts Cover a Wide Range of Uses."

5. "Data Mining: Federal Efforts Cover a Wide Range of Uses."

6. "Data Mining: Federal Efforts Cover a Wide Range of Uses."

7. Gene Healy, "Beware of Total Information Awareness," CATO Institute, 20 January 2003, available at www.cato.org/dailys/01–20–03.html, accessed 24 June 2005.

8. Theresa Hampton and Doug Thompson, "Where Big Brother Snoops on Americans 24/7," Capitol Hill Blue, 7 June 2004, available at www.capitolhill-blue.com/artman/publish/article_4648.shtml, accessed 7 February 2005.

9. Robert Jaques, "Cash under Threat from RFID Payments," VNU Business Publications, 25 February 2005, available at www.vnunet.com/articles/ print/2126822, accessed 25 March 2005.

10. Peter Clarke, "Hitachi Adds Antenna to RFID 'Mu–Chip'," EE Times, 2 September 2003, available at www.eetimes.com/story/OEG20030902S0032, accessed 18 January 2005.

11. "Security Technology: Where's the Smart Money?" Texas Instruments, 7 February 2002, available at www.ti.com/tiris/docs/news/in_the_news/ 2002/02–07–02.shtml, accessed 2 January 2005.

12. Rob Buckley, "Sense and Respond," *Infoconomy*, 1 December 2003, available at www.infoconomy.com/pages/M–iD/group87935.adp, accessed 15 June 2005.

13. "RFID Streamlines Processes, Saves Tax Dollars," Sun Microsystems website, available at www.sun.com/br/government_1216/feature_rfid.html, accessed 22 May 2005.

14. Eric Lichtblau, "Plan to Let F.B.I. Track Mail in Terrorism Inquiries," *New York Times*, 21 May 2005, available at www.nytimes.com/2005/05/21/politics/21terror.html, accessed 22 May 2005.

15. "Automatic Identification—When to Use RFID," IDF Consulting website, available at www.icfconsulting.com/Publications/Perspectives–2004/doc_files/IT–rfid.pdf, accessed 25 June 2005.

16. "ID Systems Chooses Unisys to Help Implement RFID Project for U.S. Postal Service," Unisys, 2005.

17. Ryan Singel, "No Encryption for E–Passports," *Wired News*, 24 February 2005, available at www.wired.com/news/privacy/0,1848,66686,00.html?tw=wn_story_related, accessed 26 February 2005.

18. Erik Larkin, "E–Passports Will Include New Safeguards," *PC World*, 19 May 2005, available at www.pcworld.com/news/article/0,aid,120901,00.asp, accessed 25 June 2005.

19. Alorie Gilbert, "States to Test ID Chips on Foreign Visitors," *New York Times*, 26 January 2005, available at www.nytimes.com/cnet/CNET_2100-1039_3-5552120.html, accessed 23 February 2005.

20. Paul McDougall, "Accenture's 'Virtual Border' Project," *InformationWeek*, 7 June 2004, available at www.informationweek.com/story/showArticle.jhtml?articleID=21401734, accessed 2 April 2005.

21. Paul McDougall, "Accenture's 'Virtual Border' Project."

22. "No Chip in Arm, No Shot from Gun," Associated Press, via *Wired News*, 14 April 2005, available at www.wired.com/news/technology/0,1282,63066,00.html?tw=wn_story_related, accessed 17 May 2005.

23. Curtis Lee Carrender, et al., "System and Method for Controlling Remote Devices," U.S. patent application #20020149468, assigned to U.S. Department of Energy, filed 11 April 2001.

24. "Public Affairs Plan," prepared by Fleishman–Hillard for the MIT Auto-ID Center, originally available at www.autoidcenter.org/media/public_affairs.pdf, accessed 5 July 2003. This reference has been removed from the Auto-ID Center website and is now mirrored at quintessenz.org/rfid–docs/www.auto idcenter.org/media/public_affairs.pdf.

25. "U.S. Military to Issue RFID Mandate," *RFID Journal*, 15 September 2003, available at www.rfidjournal.com/article/articleview/576/1/1/, accessed 25 June 2005.

26. Claire Swedberg, "Cattle Auctioneer Promotes Tracking Plan," *RFID Journal*, 13 June 2005, available at www.rfidjournal.com/article/articleview/1655/1/1/, accessed 26 June 2005.

27. Rick Whiting, "FDA Expects RFID Use to Combat Drug Counterfeiting," *InformationWeek*, 18 February 2004, available at www.informationweek.com/showArticle.jhtml?articleID=17701351, accessed 26 June 2005.

28. Gerry Gilmore, "Alien Touches Down: Tiny Technology Pays Off Big for N.D.," North Dakota State University, 12 July 2003, available at www.ndsu.nodak.edu/research/article.php?article_number=36, accessed 25 June 2005.

29. Florence Olsen, "Social Security Administration Utilizes RFID," *USA Today*, 5 January 2005, available at www.usatoday.com/tech/news/techpolicy/2005–01–05–rfid–to–track–ssa_x.htm, accessed 26 June 2005.

30. Jonathan Collins, "NASA Tried RFID for HAZMAT," *RFID Journal*, 14 December 2004, available at www.rfidjournal.com/article/articleview/1288/1/1, accessed 26 June 2005.

31. "ID Systems Chooses Unisys to Help . . ."

32. Mary Catherine O'Connor, "Homeland Security to Test RFID," *RFID Journal*, 28 January 2005, available at www.rfidjournal.com/article/articleview/1360/1/1/, accessed 26 June 2005.

33. G. Martin Wagner, "GSA Bulletin FMR B–7 Radio Frequency Identification (RFID)," General Services Administration website, available at www.gsa.gov/Portal/gsa/ep/contentView.do?P=MTP&contentId=17662&contentType=GSA_BASIC, accessed 26 June 2005.

34. Jonathan Collins, "Rep. Senators Vow to Protect RFID," *RFID Journal*, 10 March 2005, available at www.rfidjournal.com/article/articleview/1440/1/1/, accessed 10 March 2005.

Chapter 16 — The Nightmare Scenario

1. R.J. Rummel, *Death by Government*, Transaction Publishers, 1997, p. 1.

2. Kevin Ashton, "Kevin Ashton, Auto–ID Center, at Forrester Executive Strategy Forum," videotape, 7–9 November 2001.

3. Leonard Gross, *The Last Jews in Berlin*, Carroll & Graf, 1999, p. 28.

4. Army Reserve Pfc. Lyndie England achieved notoriety in 2004 for her role in the Abu Ghraib prison abuse scandal. England was one of several American service personnel photographed giving the thumbs up while humiliating naked Iraqi detainees. In one particularly damning photograph, England is shown detaining a prisoner on a dog leash.

5. R.J. Rummel, "20th Century Democide," available at www.hawaii.edu/powerkills/20TH.HTM, accessed 16 July 2005.

6. R.J. Rummel, *Death by Government*, Transaction Publishers, 1997, p. 9.

7. Karen Abbott, "ACLU Plans to Sue FBI over Surveillance Cases," *Denver Rocky Mountain News*, 18 May 2005, available at rockymountainnews.com/drmn/local/article/0,1299,DRMN_15_3786510,00.html, accessed 27 June 2005.

8. Michael Sawkiw, president of Ukrainian Congress Committee on America Inc., testimony to the House of Representatives Resources Committee, 2005, available at resourcescommittee.house.gov/archives/109/testimony/2005/michaelsawkiw.htm, accessed 27 June 2005.

9. See, for example: Lisa Greene, "Face Scans Match Few Suspects," *St. Petersburg Times*, 16 February 2001, available at www.sptimes.com/News/021601/TampaBay/Face_scans_match_few_.shtml, accessed 27 June 2005.

10. Hitler's quote was cited in testimony by Major Wallace at the Nuremberg trials on November 23, 1945. Major Wallis's testimony reads as follows:

    Meanwhile, during this entire pre-war period, the nation was being prepared psychologically for war, and one of the most important steps was the reshaping of the educational system so as to educate the German youth to be amenable to their will. Hitler publicly announced this purpose in November 1933, and I am quoting from Document 2455-PS. He said: 'When an opponent declares, *I will not come over to your side, and you will not get me on your side,* I calmly say, *Your child belongs to me already. A people lives forever. What are you? You will pass on. Your descendants, however, now stand in the new camps. In a short time they will know nothing else but this new community.*'

    A transcript of this hearing is available at Yale University's Avalon project, online at: www.yale.edu/lawweb/avalon/imt/proc/11-23-45.htm, accessed 27 June 2005.

11. George L. Mosse, *Nazi Culture*, Grosset & Dunlap, 1966, p. xxxviii.

Chapter 17 — Pull the Plug!

1. QuoteDB website, available at www.quotedb.com/quotes/862, accessed 24 June 2005.
2. Frederick Douglass, West India Emancipation (Aug. 4, 1857) & Dred Scott (May 1857), p. 22, in *Two Speeches by Frederick Douglass*, 1857.
3. Phyllis L. Kim, "Auto–ID Center Communications," Fleishman–Hillard for the MIT Auto–ID Center, originally available at www.autoidcenter.org/media/pk–fh.pdf, accessed 5 July 2003. This reference has been removed from the Auto-ID Center website and is now mirrored at cryptome.org/rfid/pk–fh.pdf.
4. "RFID and Consumers: Understanding Their Mindset," CapGemini and National Retail Federation, 2004, available at www.nrf.com/download/NewRFID_NRF.pdf, accessed 25 March 2005.
5. Jonathan Collins, "Consumers More RFID–Aware, Still Wary," *RFID Journal*, 8 April 2005, available at www.rfidjournal.com/article/articleview/1491/1/1/, accessed 10 April 2005.
6. Jane Black, "Shutting Shopping Bags to Prying Eyes," BusinessWeek Online, 5 March 2004, available at www.businessweek.com/technology/content/mar2004/tc2004035_8506_tc073.htm, accessed 25 June 2005.
7. Erik Larkin, "E–Passports Will Include New Safeguards," *PC World*, 19 May 2005, available at www.pcworld.com/news/article/0,aid,120901,00.asp, accessed 25 June 2005.
8. "The Nestlé Boycott," Breastfeeding.com, available at www.breastfeeding.com/advocacy/advocacy_boycott.html, accessed 28 February 2005.

Epilogue

1. Jeremy Grant, "US military 'rocks' spy world," *The Financial Times*, 27 May 2005, available at http://news.ft.com/cms/s/35bae060-ce20-11d9-9a8a-00000e2511c8.html, accessed 1 July 2005.

2. IBM press release, "IBM takes RFID to the next level," IBM, 14 June 2005, available at http://www-1.ibm.com/press/PressServletForm.wss?MenuChoice=press releases&TemplateName=ShowPressReleaseTemplate&SelectString=t1.docunid=7738&TableName=DataheadApplicationClass&SESSIONKEY=any&Window Title=Press+Release&STATUS=publish, accessed 1 July 2005.

3. United States Department of Homeland Security, "70—Radio Frequency Identification (RFID) Technology," Federal Business Opportunities website, available at http://www.fbo.gov/servlet/Documents/R/1239555, accessed 21 February 2006.

4. Catherine Komp, "Radio ID Technology Spreads; Privacy Activists Dig In," *The New Standard,* 5 May 2006, available at http://newstandardnews.net/content/?action=show_item&itemid=3129, accessed 5 May 2006.

5. Wisconsin Assembly Bill 290, "An Act to create 146.25 of the statutes; relating to: prohibiting the required implanting of a microchip in an individual and providing a penalty," introduced by Representative Marlin D. Schneider in the 2005 regular session, history available at http://www.legis.state.wi.us/2005/data/AB290hst.html, accessed 4 May 2006.

6. Catherine Komp, "Radio ID Technology Spreads; Privacy Activists Dig In."

7. WorldNetDaily, "Employees Get Microchip Implants," WorldNetDaily, 10 February 2006, available at http://www.worldnetdaily.com/news/article.asp?ARTICLE_ID=48760, accessed 10 February 2006.

8. MSN Money, "VeriChip Corporation Appoints Former Secretary of Health & Human Services and Former Governor of Wisconsin Tommy G. Thompson to its Board of Directors," MSN Money, 7 July 2005, available at http://news.money central.msn.com/ticker/article.asp?Symbol=US:ADSX&Feed=BW&Date=20050707&ID=4947241, accessed 15 July 2005.

9. Jonathan Westhues, "Demo: Cloning a VeriChip," Jonathan Westhues website, available at http://cq.cx/verichip.pl, accessed 25 January 2006.

10. Annalee Newitz, "The RFID Hacking Underground," Wired Magazine, May 2006, available at http://www.wired.com/wired/archive/14.05/rfid.html, accessed 5 May 2006.

11. VeriChip Corporation, "Form S-1 Registration Statement," website, available at http://www.sec.gov/Archives/edgar/data/1347022/000119312505250388/ds1.htm, accessed 3 January 2006.

12. Liz McIntyre, "VeriChip Photos and Forms," Spychips.com website, available at http://www.spychips.com/verichip/verichip-photos-instructions.html, posted 27 January 2006.

13. John Lettice, "Face and Fingerprints Swiped in Dutch Biometric Passport Crack," *The Register,* 30 January 2006, available at http://www.theregister.com/2006/01/30/dutch_biometric_passport_crack/, accessed 30 January 2006.

14. Melanie R. Rieback, "Is Your Cat Infected with a Computer Virus?" 2006, Vrije Universiteit Amsterdam: Amsterdam, available at http://www.rfidvirus.org/papers/percom.06.pdf, accessed 15 March 2006.

15. Tom Espiner, "RFID 'Not Safe From DoS Attacks'," Silicon.com, 18 April 2006, available at http://software.silicon.com/security/0,39024655,39158120,00.htm,

accessed 18 April 2006.
16. Jeffrey Beckman, Director of Worldwide and U.S. Communications for Levi Strauss, personal e-mail communications with Liz McIntyre.